Long Half-life

Long Half-life

The Nuclear Industry in Australia

Ian Lowe

MONASH
UNIVERSITY
PUBLISHING

Monash University Publishing
Matheson Library Annexe
40 Exhibition Walk
Monash University
Clayton, Victoria 3800, Australia
publishing.monash.edu/

Monash University Publishing brings to the world publications which advance the best traditions of humane and enlightened thought.

ISBN: 9781922464491 (paperback)
ISBN: 9781922464507 (pdf)
ISBN: 9781922464514 (epub)

Design: Les Thomas
Typesetting: Jo Mullins

Cover image: Freeze motion of yellow dust explosion isolated on black background, by Kitsana1980. Shutterstock image 1456964321.

A catalogue record for this book is available from the National Library of Australia.

CONTENTS

INTRODUCTION

Radioactive materials break down and release radiation. Half-life is a measure of how long it takes for one of these materials to be reduced to half its original mass. Technetium, which is commonly used for medical imaging, has a half-life of about six hours, so it rapidly decays in the body. The most common form of uranium has a half-life of nearly 4.5 billion years, about as long as the Earth exists. That is why uranium poses a fundamental challenge to politicians and officials, whose time horizon rarely extends beyond this year's budget or next year's election.

An analysis of the social and political history of Australia's role in the nuclear industry inevitably illuminates general questions about the role of technology in modern society, the gradual erosion of trust in both technical experts and political institutions, the existential threats to our civilisation posed by climate change and nuclear weapons, and the fundamental incapacity of most elected politicians to engage with long-term issues. When I told a respected colleague I was writing a book about this, I was asked 'Why are you bothering?' It was a good question.

Most people had decided years ago, if not decades ago, that there was no place in Australia for nuclear energy. The fundamental problems of radioactive waste management and the link with nuclear weapons remain unresolved. The economics were always problematic, even when the last serious proposal for a nuclear power station in Australia was being canvassed fifty years ago, and the trend has worsened. Even the most creative accounting cannot get the cost of nuclear power anywhere near that of large-scale solar and wind with enough storage to be available around the clock.

I still think there are some general lessons from our experience in the nuclear era, which coincides almost exactly with my lifetime. Just to underline the fact that those lessons can easily be forgotten or overlooked, as I was writing an organisation called Future Directions International published a discussion paper that said 'the creation of a nuclear power sector ought to be revisited'. Future Directions is an independent research organisation set up by a retired military officer who went on to become Governor-General, the late Major-General Michael Jeffery. The organisation has done very important work about the maintenance of Australia's soils and the productivity of agricultural land. This new paper argued that a nuclear power industry would enable Australia to operate a nuclear-powered submarine fleet and said it was attempting 'to restart the discussion of the advantages' of having an expanded nuclear industry.[1] I will return to this issue towards the end of this book.

When uranium and radium were first discovered in Australia more than a hundred years ago, those elements were seen mainly to have fringe applications in medicine. The Australian physicist Mark Oliphant was a member of the famous Cambridge research group led by Lord Rutherford, whose developing understanding of the basic physics of radioactive elements led to the Manhattan Project and the atomic bombs that ended World War II.[2] When the United States decided not to share the new weapons with its wartime allies, Australian governments supported the development of the British bomb by supplying uranium from relatively small mines and by allowing the weapons to be tested on Australian land. The public were led to believe that the same principles could be applied to generate electricity, with our uranium being used to provide clean energy. The Australian Atomic Energy Commission (AAEC) was established and funded to build a prototype nuclear reactor in what was then bushland well outside the suburbs of Sydney. It was envisaged that the AAEC program at Lucas Heights would develop the expertise to allow Australia to build and operate nuclear power stations. It became apparent later that there was political interest in the possibility that we could also produce nuclear weapons.

Various state governments expressed interest in having nuclear power stations, with prominent scientists Sir Philip Baxter and Sir Ernest Titterton confidently predicting that the new technology would replace coal-fired electricity.[3] When this failed to happen, those scientists lamented the unwillingness of politicians to accept their expert advice and argued that politically motivated activists were spreading misinformation to confuse the public. The short-lived Gorton Liberal Government was persuaded to plan a nuclear power station on land owned by the Commonwealth Government at Jervis Bay, but that project was first postponed and then abandoned by later administrations. It has more recently been revealed that there were secret negotiations aimed at expanding Australia's role in the nuclear industry by subsidising uranium enrichment facilities, while Baxter and Titterton also made persistent public calls for nuclear weapons to be developed as part of our defence strategy.[4]

The discovery of large uranium deposits in the Northern Territory in the 1970s raised the possibility that Australia could become a major exporter of radioactive minerals. The Fox Report, commissioned by the Whitlam Labor Government to study the environmental impacts of the proposed Ranger uranium mine in the Kakadu region of the Northern Territory, broadened into an inquiry into the social and political issues of preventing weapons proliferation and managing long-lived waste from nuclear reactors.[5] For a brief period these issues were politically significant, with the Australian Labor Party (ALP) opposed to exporting uranium and the Fraser Liberal Government supportive. In the early 1980s Bob Hawke persuaded the ALP to modify its opposition and support the Olympic Dam project in South Australia, a major copper mine at Roxby Downs, which also produced gold, silver and uranium. Opposition from Indigenous groups caused the proposed Jabiluka mine to be abandoned, but for a few decades there was bi-partisan support for export of uranium from the large Ranger and Olympic Dam mines. Public concern about the broader issues did not go away and for a brief period during the Cold War, when there was the constant threat of global nuclear war, the Nuclear Disarmament Party played a significant role in Australian politics.

At the international level, the 1986 Chernobyl accident seriously damaged the reputation of the nuclear industry for providing clean energy. For a brief period after the United Nations (UN) climate change treaty was developed in the early 1990s, advocates of nuclear energy made some headway with their argument that it could be part of the necessary decarbonising of the electricity system, but the Fukushima disaster and worsening economics have seen most countries abandon plans to expand nuclear power. Within Australia, Liberal Prime Minister John Howard set up an inquiry when his inaction on climate change became a political liability, but the Switkowski report showed that nuclear power was neither a timely nor a cost-effective way to reduce Australia's greenhouse gas emissions.[6]

Radioactive waste has been a persistent political issue in Australia. The Lucas Heights reactor produces wastes that need to be managed, while medical and industrial applications of nuclear technology also produce so-called low-level waste, mildly radioactive materials that need to be stored safely. For decades, successive Commonwealth governments have tried unsuccessfully to get approval for construction of a national facility to store these low-level wastes. While I was writing, the chief executive of the Committee for Adelaide publicly lamented the fact that government dithering had resulted in low-level waste continuing to be stored in the centre of that city, rather than somewhere in the outback. There is also a continuing headache in the form of the so-called Intermediate Level Waste from the Lucas Heights establishment. The New South Wales Government is unhappy about the waste remaining at that site, which is now surrounded by the suburbs of the Shire of Sutherland, but there is currently no alternative.

More recently there have been suggestions that Australia might offer to store and dispose of the high-level wastes from nuclear reactors in other countries. The Weatherill Government in South Australia was sufficiently interested in the idea to set up and fund a Royal Commission, which concluded that there would indeed be a significant economic opportunity in managing the waste from nuclear power stations in Japan, South Korea and Taiwan.[7] The commission observed that such a major investment would require clear public support, so the government established a citizens' jury to consider the issues; to the surprise of the elected politicians, the citizens strongly recommended against the proposal, which was subsequently dropped. The issue raises important questions about trust in experts and trust in governments to manage complex long-term issues.

There are other contemporary questions. Many in the defence industry believe that the next generation of submarines should be nuclear powered, while the increasing strength of China and the perception that the United States might not be willing to act as our protector has led to renewed calls to discuss the issue of nuclear weapons. Two huge problems persist at the international level. Large volumes of dangerous 'high-level waste' are still being stored near nuclear power stations, with most of the countries involved having no clear plan to manage those materials for the unimaginably long periods required. The Nuclear Non-Proliferation Treaty has not led to weapons-holding nations disarming, so predictably others have joined the nuclear club, with the real risk that tensions in the Korean peninsula, the Indian subcontinent or the Middle East could lead to the catastrophic consequences of nuclear weapons being used against cities. Concern about the moral responsibility for the uses of Australian uranium raises the question of the social licence to continue exporting. At different times throughout the last sixty years, politicians in Australia and other countries have used nuclear issues opportunistically in the hope of gaining short-term advantage. This inevitably raises the worry that politicians and other decision-makers, almost inevitably having a short-time horizon, might be incapable of making the decisions needed to protect civilisation from the existential threats of climate change and nuclear war.

Since an important theme here is the way we perceive reality through the lenses of our experience and our values, I want to set out clearly where I stand. I studied electrical engineering and science as a part-time student at the University of New South Wales, while earning a living designing and building electronic equipment. Growing up in that state, I saw the pollution from coal-fired power stations and I also heard regularly of accidents in coal mines. So I was quite enthusiastic about the promise of cleaner and safer energy from nuclear power stations. After I graduated with honours in physics, I taught school science while I thought about what sort of life I wanted to lead. Having decided to pursue an academic career, I accepted a scholarship at the University of York to undertake doctoral research. My project was funded by the UK Atomic Energy Authority, studying the basic physics of a problem that affects the useful life of fuel elements in nuclear reactors. The Wilson Government had been elected in the United Kingdom in 1964 on a platform of supporting technological change, with its enthusiasm for nuclear energy a central component.

My first academic appointment in 1971 was a lectureship in materials science within the Faculty of Technology at the newly established UK Open University (OU). Some of my colleagues in that faculty were quite critical of the economics and the safety of nuclear power stations, forcing me to review my thinking and temper my enthusiasm for that approach. As one of the founding members of the Energy Research Group at the OU, I was involved in studies of the United Kingdom's future energy options, including one project that revealed a proposed crash program to build thirty-six nuclear reactors would require so much conventional energy it would actually create the energy crisis it was aiming to resolve.

In 1977, I spent six months as a visiting fellow at Griffith University and became involved in the public discussion of the Fox Report, the outcome of the public inquiry about the proposed Ranger uranium mine in the Kakadu region of the Northern Territory. At the time, I would have considered myself a critical friend of the nuclear industry, so I was disappointed to find that elected politicians and mining company leaders were actually lying to the community about several aspects of nuclear energy: the economics, the operating risks and the big problems of radioactive waste management and weapons proliferation. I felt obliged to correct some of the untruths. In the simplistic world of political debate, that saw me rebadged as an opponent of nuclear power.

I was certainly critical of the enthusiasm for exporting uranium, as I accepted the argument of the Fox Report that this would contribute to two big problems. Nuclear power stations inevitably produce extremely dangerous waste products, for which there was no effective management scheme, despite the assurances of Liberal Party politicians and mining company executives. Any country with nuclear power stations could potentially develop nuclear weapons, as India had demonstrated in 1974.

I returned to Australia permanently in 1980, taking up an appointment at Griffith University, where I lectured in the broad field of science, technology and society as well as directing the Science Policy Research Centre. I researched Australia's future energy options and was appointed to the National Energy Research, Development and Demonstration Council, chairing its panel on economic, social and environmental issues. In that capacity, I was involved in early discussions about the emerging scientific evidence that use of fossil fuels was changing the global climate. That led to other advisory roles, including directing the Commission for the Future in 1988 and chairing the council that produced in 1996 the first independent national report on the state of the environment. As a member of the Radiation Health and Safety Advisory Council for fourteen years, I was involved in discussions with the regulatory body, the Australian Radiation Protection and Nuclear Safety Agency, about the entire spectrum of issues arising from radiation. In 2006, I became concerned about suggestions that we might build nuclear power stations as a response to climate change, so I gave a presentation to the National Press Club and subsequently wrote a Quarterly Essay on the subject.[8]

In 2010 I collaborated with Professor Barry Brook to write a 'flip book' about nuclear power.[9] As part of a series released by a small Sydney publisher, Pantera Press, the book had no back cover but two alternative front covers. One way up it said *Nuclear Power – Yes*, with an essay by Professor Brook saying why he thought Australia should build nuclear power stations followed by my critique of his argument. Turned over, the book said *Nuclear Power – No*, with my essay saying why we should not build nuclear power stations followed by Professor Brook's critique. We subsequently did a few double-acts at ideas festivals and writers' festivals to explain our respective views. At one of these, a member of the public asked how two intelligent and well-informed scientists could come to such completely different conclusions. I explained that we both accepted the science of climate change and the consequent need to move away from burning coal to produce our electricity. The only credible large-scale alternatives were nuclear energy or renewables, solar and wind. We both accepted the limitations of the existing nuclear power stations and their contribution to the big problems of waste management and weapons proliferation. In analysing the feasibility of the future options, our underlying values had caused us to reach the different conclusions that were on display. Professor Brook was confident that new designs of nuclear reactors could solve those problems and provide virtually unlimited clean energy, while he was sceptical about the possibility of scaling up solar and wind to provide the massive amounts of energy needed by a modern society. By contrast, I was confident that our huge land area meant we could develop enough solar and wind energy to power the country, but I was not impressed by the promise of a new generation of nuclear reactors.

These were both intellectually defensible interpretations of the uncertain future we faced in 2010. I think what has happened since then has strengthened my case. South Australia got the majority of its electricity from solar and wind in 2020, on some occasions being entirely powered by the renewable energy technologies, and the governments in New South Wales and South Australia have both set out commitments to being almost completely dependent on solar and wind by 2030. Both those states have Liberal–National Coalition administrations. By contrast, there has been no significant progress in developing and deploying either Generation Four reactors or small modular reactors. That's because there remains a fundamental contradiction in the argument for building nuclear power stations in Australia. Every competent analyst agrees that the current designs cannot compete economically. The only way of making a case for future nuclear power stations being economically viable is to assume that others will build the new designs that are still on the drawing board and that a learning curve will significantly improve the economics. It is clear, however, that we need to reduce the greenhouse gas emissions from power stations urgently to play our part in slowing global climate change. That means the economics are unacceptable for nuclear power to be a timely answer, while the only way of making the economics look acceptable is to live with an unreasonable delay in cleaning up our electricity system.

The differences between my conclusions and those of Professor Brook were a reminder that we cannot be objective when we look at the complex modern world and consider alternative solutions for the problems we face. We all see the world through the lenses of our experience and our values. If you talk to supporters of the two opposing teams after a closely contested football match, it can be difficult to believe they actually saw the same events. They will disagree about matters of fact, such as whether the ball went out of play or a collision between opposing players was deliberate. They will disagree about matters of opinion, such as whether a free kick should have been awarded. They will often disagree about the officials' eyesight or their impartiality.

They are seeing what they want to see. I have noticed that dedicated supporters are almost always optimistic about the chances of their team having a good season, even if the players are mostly the same group that did poorly the year before. While science has been described as 'the true exemplar of authentic knowledge', implying that scientists are objective and all competent professionals will reach the same conclusion: it is only true of controlled measurements in a laboratory. Ask for a measurement of the melting point of lead or the tensile strength of a specified alloy and you can expect a reliable answer from any competent scientist. However, the US nuclear scientist Alvin Weinberg pointed out in 1962 that there is a class of problems he called 'trans-science'. The two specific examples he cited are both relevant to the nuclear industry: the health risks of low levels of ionising radiation and the operating safety of nuclear reactors. Weinberg argued that these are clearly scientific questions, they are posed in the language of science, but there is no possibility even in principle of giving an 'objective' answer that would be accepted by any competent professional.[10] Those problems still defy comprehensive analysis.

The objectivity of nuclear scientists was first questioned in Australia in the Fox Report.[11] The report noted that many exaggerated statements had been made about the risks of uranium mining and the nuclear industry more broadly in public submissions. The authors of the report went on to express surprise at the lack of objectivity evident in the statements of some of those who supported uranium mining and development of nuclear power, 'including distinguished scientists'. I discuss later some of those statements, but I have since encountered an extreme example of this lack of objectivity.

As already mentioned, my doctoral research was funded by the UK Atomic Energy Authority (UKAEA), in particular by a group involved in the development of the prototype Fast Breeder Reactor. This was an innovative design, called a breeder because the aim was not to use up nuclear fuel like uranium but use it to breed the nuclear materials for use in future reactors. The prototype reactor was situated at Dounreay, on the very northern extremity of the Scottish mainland. It was as far away from large settlements as it could possibly be. David Collingridge, in his seminal book *The Social Control of Technology*, quoted the reasoning put forward by Lord Hinton, the head of the UKAEA at the time of the project.[12] He accepted that the reason for siting the reactor in a remote site was the concern that radiation would leak out of the containment vessel and that poses a risk to nearby residents. 'So we assumed, generously, that there would be one per cent leakage,' he said. When they calculated the risk to the relatively small number of people living in that remote region, they found that 'the site did not comply with the safety distances specified by the health physicists'. What did they do? Did they revise the design of the reactor, or move to an even more remote site? No. In Hinton's own words, quoted in Collingridge's book, 'That was easily put right; with the assumption of a 99% containment the site was unsatisfactory so we assumed, more realistically, a 99.9% containment and by doing this we established that the site was perfect'.

This is a startling statement. It does not concede that if the more generous assumption were valid, the site would have been acceptable; it asserts that making the new assumption, essentially dividing the radiation dose to the affected community arbitrarily by a factor of ten, *established* that the site was perfect!

The world is living through the aftermath of the Trump presidency in the United States and the rebadging by his office of barefaced lies as 'alternative facts'. When we discuss complex issues that pose an existential threat to civilisation, as global climate change and nuclear weapons do, it is important for the debate to be based on solid facts. It is customary for skilled politicians to promote their arguments by making statements that are factually correct but deliberately misleading. There is a reason that courts ask witnesses to swear to tell 'the truth, the whole truth and nothing but the truth'. That formula recognises that telling only part of the truth, or mixing true statements with others which are not true, can be misleading. Until recently, most politicians were careful only to make statements that were misleading and avoid the demonstrably false, but we seem to have moved recently into dangerous territory where leaders feel comfortable lying to advance their political argument.

Australian Prime Minister John Howard told voters during the 2001 election campaign that desperate asylum seekers had thrown their children overboard, then joined US President George W. Bush and UK Prime Minister Tony Blair in claiming Iraqi leader Saddam Hussein had weapons of mass destruction to justify an invasion of that country. Those were just lies. Before the 2013 election, Tony Abbott promised that if he were elected there would be no cuts to the ABC or SBS, but his 2014 budget savaged public broadcasting. More recently, the major Australian political parties have based entire election campaigns around dishonest portrayals of their opponents, the ALP in 2016 claiming that the Liberal–National Coalition intended to privatise Medicare and the Coalition's supporters in 2019 claiming that the ALP was planning to introduce what they called 'a death tax'. Asked in 2021 why the ABC was facing more budget cuts, Prime Minister Scott Morrison looked straight at the questioner and asserted that the ABC's budget was actually increasing. 'There are no cuts,' he said with a straight face. It reminded me that the late Clive James, when he was TV critic for a British newspaper, said that the characteristic of a successful politician was to look sincere while lying through their teeth. The filmmaker Woody Allen made a similar comment when he said that most important virtues for a politician were sincerity and integrity. 'Once you can fake those, you've got it made,' he said.

With the Murdoch press totally unreliable and the public broadcasters being systematically deprived of the resources needed for critical journalism, it is very difficult for the public to have confidence they are being truthfully informed about complex issues. Just as misinformation was used for decades by the tobacco industry to diffuse support for measures to control their deadly product, a deliberate campaign of misinformation has been used to prevent implementation of the obvious cost-effective means to slow down global climate change. Future generations will probably be bemused by the way successive governments have adopted what a British TV comedy called 'masterly inaction' in the face of these serious issues. If they scrutinise the discussion of nuclear issues in Australia, they will be equally puzzled by two features. Firstly, important and irreversible decisions about these issues have often been made without any attempt to obtain the informed consent of the community. Secondly, where there has been public debate, misinformation and political opportunism have often prevailed.

While I have agreed with some of the decisions and disagreed with others, I have consistently been uncomfortable about the way the choices were made. I agree with a statement made in the introductory section of the Fox Report:

> Ultimately, when the matters of fact are resolved, many of the questions which arise are social and ethical ones. We agree strongly with the view, repeatedly put to us by opponents of

nuclear development, that, given a sufficient understanding of the science and technology involved, the final decisions should rest with the ordinary man [*sic*] and not be regarded as the preserve of any group of scientists or experts, however distinguished.[13]

The unconscious sexism of its time aside, this is a critical point.

In putting the recommendations of its Nuclear Fuel Cycle Royal Commission to a citizens' jury, the South Australian Government followed that principle. Scientific experts can tell us how to store radioactive waste safely, while engineers can tell us how to construct the storage systems. But deciding whether we want to store radioactive waste from other countries as a commercial operation is an ethical question for society as a whole, not one for technical experts. The role of the experts was to brief the citizens and answer their questions, admitting uncertainty where it exists, to provide an informed basis for the jury to debate the social and ethical questions. The jurors had to weigh up the various estimates of the possible economic benefits of storing radioactive waste against the differing views about the environmental and financial risks.

An important zoom discussion took place while I was writing this book, when face-to-face meetings were precluded in order to slow the spread of the coronavirus. The 25 leading Australian thinkers of that discussion concluded that our society faces two serious existential threats: climate change and nuclear weapons.[14] In both cases, the wrong decisions could lead to the collapse of civilisation.

I have written elsewhere about climate change. It is relevant to this book only in so far as nuclear energy can be regarded as potentially part of the elimination of fossil fuels from the electricity system. The issue of nuclear weapons is central here, for a reason. The first public demonstration of nuclear science was the bombing of Hiroshima and

Nagasaki in 1945 by the US Air Force. Any country that uses nuclear science to produce electricity has the basic infrastructure to produce nuclear weapons, as has been demonstrated by the nations that have done so. The more countries have nuclear weapons, the more likely it becomes that a deranged or desperate leader will use them, or that a failure of communication could result in their inadvertent use. An understanding of both the basic science of nuclear physics and its engineering applications is an essential framework for discussing these troubling issues. So is a grasp of the political questions that have determined how the nuclear industry has evolved, both in Australia and globally.

Chapter 1

THE DAWN OF THE NUCLEAR AGE

In the late nineteenth century, the distinguished British scientist Lord Kelvin famously said that almost all the important problems in physics had been solved. There only remained, he stated, an explanation of the photoelectric effect – that shining light on some substances could produce an electric current – and the negative outcome of the Michelson–Morley experiment, a famous attempt to measure the speed of light. It is now part of the history of science that a young Albert Einstein, at the time working in the Swiss patents office, solved both those problems with papers published in 1905. Rather than completing the work of physics and providing a complete understanding of the natural world, Einstein's insights opened up entirely new fields of research, quantum physics and relativity, that eventually combined to produce nuclear weapons. Let me start by explaining the two bits of basic physics mentioned above.

First, the photoelectric effect. In 1900, it was accepted that light behaved like waves. This model explained everyday effects like reflection from a mirror and refraction – the bending of light by glass, used for a range of practical devices from spectacles to telescopes – as well as more complex effects like interference, when beams of light interact just as waves do in water. The splitting of a beam of white light by a glass prism was understood to show that the various colours of light

17

each have a specific wavelength and so they are affected differently by the transition from air to a denser medium like glass or water. But the photoelectric effect was a puzzle.

It had been observed that you could produce an electric current by shining light onto a metal surface. By this time it was known that the current was a flow of small particles called electrons, so the light was releasing those electrons from the metal surfaces. The unexplained problem was that the effect seemed to be dependent on the colour of the light – in technical terms, its wavelength or frequency, rather than the intensity. For some metals, no electrons were released by red light, no matter how intense the beam, but even weak beams of green or blue light produced a current.

Einstein's insight was to recognise that the light was behaving as if it were a stream of particles, with their energy proportional to the colour or frequency. He showed that these photons could, if sufficiently energetic, release electrons from the metal, with each type of metal having a characteristic energy level needed for this to happen. He linked this understanding to earlier work by German theoretical physicist Max Planck, explaining the radiation of light from heated surfaces, showing that the energy of photons was related to their frequency by a number, now known as Planck's constant. We routinely talk about a substance being red-hot or white-hot, recognising that the light coming from a surface is dependent on its temperature. The fixed amount of energy possessed by a photon of particular colour became known as its quantum, and the whole field of quantum physics developed from that insight. The idea has passed into everyday language, with people talking about the quantum of a pay increase or making a quantum leap from one position to another.

The Michelson–Morley experiment was an attempt by American physicists Albert Michelson and Edward Morley to measure how the speed of light was affected by the direction of travel. Since light was known to behave like waves, physicists believed that there had to be a medium for the light to travel in. It had become accepted that luminiferous ether was this medium. We now know this was a misconception. Light can travel through a vacuum, as it does when it travels through space from the sun or a distant star. The experiment set out to measure how the speed of light was affected by the rotation of the Earth. It was based on the known fact that the apparent speed of a moving object can be affected by any relative motion of an observer. For example, if a train was travelling through a station at 30 kilometres per hour and a standing passenger threw a ball along the carriage, the speed as seen by another person in the train would differ from the speed observed by someone standing on the platform. If the ball was thrown at 20 kilometres per hour in the same direction the train was moving, the observer on the station platform would see the ball moving at 50 kilometres per hour relative to the station. If the ball had been thrown in the opposite direction, the person on the platform would have seen it only moving at 10 kilometres per hour. The Michelson–Morley experiment split a beam of light and sent the two halves in different directions, expecting to measure a difference. They found none.

Einstein's explanation was counter-intuitive. He argued that the result must mean that the speed of light is not dependent on which way it was travelling relative to the Earth. From this starting point, he developed what became known as the Special Theory of Relativity. Its most important conclusion, for the purpose of this book, was that it established a relationship between mass and energy.

One of the most fundamental principles of physics had been the principle of conservation of energy, which held that energy could not be created or destroyed, only changed from one form to another. Lifting an object gives it potential energy, dropping it turns that into kinetic energy. Passing an electric current through the filament of a light bulb transforms electrical energy into heat and light. Einstein showed that accelerating an object to a very high speed could increase its mass, and inversely mass could be turned into energy, the two being related by what is probably his most famous equation, $E = mc^2$. Energy is equal to mass multiplied by the square of c, the speed of light. Since the speed of light is very fast, about 300,000 kilometres per second, we usually see it as instantaneous and it becomes significant only over very large distances. Light from the sun takes about eight minutes to reach the Earth. In the standard International System of Units (SI – from the French *Système international [d'unités])*, the velocity of light is expressed as 300 million metres per second, so this number multiplied by itself is huge: 90 billion million, or 9 with 16 zeros after it. The significance of this became apparent several decades later, when scientists realised that splitting the nucleus of a heavy atom like uranium gave two lighter nuclei with slightly less mass, releasing a large amount of energy.

At the time, nuclear science did not exist because the structure of atoms had not yet been figured out. But radioactive materials had been known in the nineteenth century. The French physicist Henri Becquerel discovered radioactivity in 1896 and received the 1903 Nobel Prize for physics along with the Curies – Pierre and Marie – who had done important research to analyse the properties of radioactive materials. Becquerel found that a lump of uranium clouded a photographic plate and concluded that it was giving off some sort of rays. The names of those founders of nuclear science are preserved in the units used to indicate the level of activity from a radioactive substance. The becquerel, the internationally used unit, is equivalent to one decay per second. That means it is a very small unit, with the radiation from commonly occurring substances often millions or billions of becquerel. In the United States it is still common for the earlier unit, the curie, to be used. The conversion is that a curie is roughly 37 billion becquerels.

Studying radioactive materials, the Curies identified polonium and radium. Pierre was killed in a road accident in 1906, but Marie continued the work they had begun together, becoming the first woman appointed to teach at the Sorbonne and was awarded a second Nobel Prize, for chemistry, in 1911. At the time, the effects were simply a scientific curiosity. Uranium had been discovered in Europe in the eighteenth century, when it was named after the planet Uranus, and deposits were identified in Australia around 1900. Radium was intermittently mined between 1906 and 1934 from the South Australian site named Radium Hill by young geologist Douglas Mawson, who would go on to explore the Antarctic. As Alice Cawte wrote in her excellent book *Atomic Australia*, radium was used in early treatments of cancer and for 'health and beauty treatments purporting to cure all things from radium to "double chin".'[1] Uranium was mostly seen as a waste product, but it was used in paints and ceramics. Cawte noted that the yellow tiles that were a feature of the façade of the Mark Foys department store in the Sydney CBD were produced using uranium from Radium Hill. That building now houses law courts.

I am indebted to Alan Melbourne of Australian Radiation Protection and Nuclear Safety Agency (ARPANSA), whose 2009 presentation on the history of radium I still treasure. It included a summary of what he called 'unusual historical uses of radium'.[2] Some of them are truly startling. Ceramic jars containing drinking water with radon added were popular in the early twentieth century. Radium bread was made by the Hippman-Blach Bakery in what is now the Czech Republic. Tho-radia face cream containing thorium chloride and radium bromide was sold in France in the 1930s. Radium chocolate was marketed in Germany in the 1930s, with advertising claiming powers of rejuvenation. Radium suppositories claimed to be 'for restoring sex power' were sold by a company in Denver, Colorado, offering 'plain packaging for confidentiality'. Radium toothpaste was produced in Germany during World War II, the manufacturer claiming that it increased the defences of the teeth and gums, 'polishes the dental enamel and turns it white and shiny'.

Radium mud baths at a site in Turkey were claimed to be 'good for the treatment of rheumatic, dermatological and gynaecological diseases as well as neurological and physical exhaustion'. Bailey Radium Laboratories added radium to distilled water and marketed it as 'Perpetual Sunshine', claiming that it would cure stomach cancer and mental illness as well as restoring 'sexual vigour and activity'. Mr Melbourne added that the product gained notoriety when Eben Byers, a prominent industrialist who drank a bottle a day for four years, died as cancer of the jaw caused his facial bones to disintegrate. Along similar lines, a sad fate befell the inventor of the Scrotal Radioendocrinator. The device consisted of seven radium-soaked pieces of paper, covered with thin clear plastic and two gold wire screens. Men were advised to place the instrument under the scrotum at night like an 'athletic strap', Mr Melbourne says, with the promise it would 'invigorate sexual virility'. The inventor later died of bladder cancer, presumably from the use of his weird device.

The sad cases of radium suppositories, Perpetual Sunshine and the Scrotal Radioendocrinator illustrate the susceptibility of ageing men to a sales pitch promising to restore their flagging virility. As a man who reveals his age by playing over-70s cricket, I regularly receive email messages urging me to worry about my declining testosterone levels and promising magic cures to revitalise me, in plain packaging so as to conceal my gullibility from the postal system.

A significant workplace health issue arose in the 1920s from the practise of using radium paint for the luminous dials of watches.[3] Women sat in rows with radium-laced paint on their desks and were encouraged to lick their paintbrushes to allow fine painting of the watch dials. Suspicions arose when painters were found to be developing lesions on their jaws. Cancers were reported as early as 1924, but it took until 1929 for the cancers to be proven results of the radium paint. While these examples of historic uses of radium all seem now to reflect a scarcely credible level of ignorance, I recall that as recently as the 1980s, a popular brand of mineral water being sold in Queensland boasted a significant level of radioactivity. It has since been withdrawn from sale.

Charles Pope, a friend from my student days, has written that the extraction of uranium left a significant environmental legacy in an inner suburb of Sydney.[4] A smelter operated from 1911 to 1915 in Hunters Hill, mostly known when I lived in Sydney as the site of a famous private school. It extracted radium from uranium ore and left the tailings on the site. Some remediation took place in the 1960s, but concerned residents sparked a parliamentary inquiry in 2008. Perhaps predictably, it found radiation levels that Pope says were 'several times background levels' in parts of the site.

At the start of the twentieth century, our understanding of the structure of atoms was very primitive. It was known that they contained positive particles, which were called protons, and much smaller negative particles called electrons. As Pope wrote in his book *Living with Radiation*, the accepted picture was called the plum pudding model, 'a positive sphere (the pudding) containing negatively charged particles (the plums)'.[5] The New Zealand physicist Ernest Rutherford conducted important experiments that showed atoms appear to be mostly empty space. As Pope explains it, the model Rutherford developed for the simplest atom, hydrogen, could be illustrated by imagining a sphere the size of the Melbourne Cricket Ground (MCG) with the nucleus as an object the size of a soccer ball in the centre circle and the single electron as a bee buzzing around in the space around it. That model explains why Rutherford was able to fire protons at a metal foil and have them pass straight through. A soccer ball fired across the MCG would be very unlikely to strike the buzzing bee. The larger atoms of heavier elements have a larger nucleus, which experiments in Rutherford's Cambridge laboratory showed to contain neutral particles he called neutrons as well as protons, and a cloud of electrons, but the atoms still behave as if mostly space.

The elements the Curies had studied were called radioactive because they released radiation. What became known as nuclear science was literally the science of the atomic nucleus. The research being done in the 1930s established that the radiation was released when heavy atoms such as uranium spontaneously broke down into lighter elements, releasing energy according to Einstein's equation. It became understood that there were natural decay chains. The story is complicated because most elements occur naturally in different forms known as isotopes. Uranium atoms will always have 92 protons and 92 electrons, but they can have different numbers of neutrons; the most common form, uranium-238, has 146 neutrons. It is comparatively stable, but when an atom of uranium-238 emits radiation it turns into an atom of thorium-234. The rate of a process such as this is characterised by its half-life – the time it will take for half of the atoms to turn into the lighter form. In the case of uranium-238, the half-life is about 4.5 billion years. That is about how long this planet has existed, so the amount of uranium is now about half what was here originally. Thorium-234 is itself much more radioactive, turning into a lighter element called protoactinium with a half-life of about 24 days. That atom is even more unstable than thorium-234, with a half-life of less than 7 hours, turning into a lighter isotope of uranium, uranium-234. You can do the arithmetic, knowing it has 92 protons and 142 neutrons. The process goes on through another eleven steps before it finishes up as a stable atom, lead-206.

When these phenomena were being studied in the 1930s, they were simply scientific curiosities. That changed soon after the outbreak of World War II. The Australian physicist Mark Oliphant had gone to Cambridge and worked in Rutherford's famous research team. His career is the subject of an engaging biography by Stewart Cockburn and David Ellyard.[6] Oliphant left Cambridge to be professor of physics at the University of Birmingham, where he supervised the research of British nuclear physicist Ernest Titterton, a recurring figure in later chapters. Oliphant was also involved in the drafting of what became a famous memorandum by expatriate German physicists Otto Frisch and Rudolf Pieirls.[7] It identified the possibility of what they called a chain reaction. They showed that, if the decay of a radioactive element released an energetic particle it could in turn cause the decay of another atom, releasing another energetic particle. This could then cause the decay of another atom, and so on. Under the right conditions, they argued, this could be the basis for a new and powerful weapon: an atomic bomb. The term 'critical mass', the amount of a radioactive substance necessary for the reaction to continue, has since passed into the language. In physics, it means the minimum amount of fissile material needed to maintain a nuclear chain reaction; in common parlance, it now means the minimum size or amount of resources needed for a venture to be viable.

The early work studying radiation has led to an understanding that radiation is all around us. Scientists talk about background radiation, to which we are exposed from the rocks beneath our feet, from space, from the buildings we live and work in, and even other humans. The level of background radiation varies significantly from place to place, higher where there are rocks such as granite under the ground, typically lower in sandy estuaries. The average dose of radiation we all receive is about 2.6 milliSieverts (mSv) per year, but some places have lower readings and others higher. Within Europe, for example, the average dose in England is only 2 but in Finland it is about 7. In the booklet *Nuclear Electricity*, Ian Hore-Lacy and Ron Hubery give figures for various parts of Australia.[8] Radiation from the ground in Sydney is less than 1 mSv a year, but in Armidale it is about 2.5; around Perth, there is a wide variation from 0.02 to 3 mSv a year. They also note the impact of altitude, saying that aircrew can receive up to 6.5 mSv a year from the time they spend at high altitude, exposed to the radiation reaching the Earth from space.

The health effects of these relatively small doses of radiation is an example of the category of problems discussed in the introduction: trans-science. Alvin Weinberg observed that the relationship between large doses of radiation, such as those suffered by people living in or near Hiroshima in Japan when the first nuclear weapon was dropped, and health outcomes is linear: twice the dose, twice the probability of damage. At the lower levels of background radiation, there are no data showing a systematic difference between, for example, the English and the Finns. Any attempt at finding a difference would be overwhelmed by the variations in diet, use of tobacco and alcohol, patterns of work, age distribution, and so on. If the pattern observed at high doses is extended downward, it is consistent with one possible theory, that the risk of damage is directly proportional to the dose of radiation at all levels.

Because no hard data are available for low doses, there is an alternative theory, that there is a threshold below which radiation does no harm. It is not possible to prove which theory is correct. Clearly, controlled experiments would not be ethically acceptable, and Weinberg suggested they would be practically impossible even if they were ethically acceptable.[9] Since there is no way of proving which theory is correct, there is an inevitable tendency for people to choose the one they would prefer to be right. Most environmental activists believe there is no safe dose of radiation, whereas many scientists who work in the nuclear industry and are exposed to extra small doses of radiation believe there is a threshold below which it does no harm. The UN Scientific Committee on the Effects of Atomic Radiation (UNSCEAR) leans towards the Linear No Threshold or LNT theory, a view that there is no threshold below which radiation is harmless.[10] That certainly reflects the cautious approach that it is better to avoid a risk than find out later people have been needlessly harmed.

As data have accumulated, standards for occupational exposure have gradually tightened. Rosalie Bertell gave data for permissible annual workplace exposure in the United States.[11] Translating into the modern units, in 1925 the level was set at 520 milliSieverts. In 1934, it was reduced to 360 mSv, then to 150 in 1949 and 50–120 in 1959.

The current permissible occupational exposure in Australia is set by the regulator ARPANSA at 20 mSv per year as an average over five years, with no more than 50 mSv in any one year.[12] So the permissible radiation dose that can be received in the workplace is about eight times the normal average background radiation.

Pope argues that these sorts of radiation doses have only minor health impacts.[13] He says that 40 per cent of Australians will contract some form of cancer and half of those will die as a result. Using the linear model, he estimates that a dose of 10 mSv would increase the risk of dying from cancer from 20 per cent to 20.05 per cent, whereas a dose of 100 mSv would increase it to 20.5 per cent. So radiation has a measurable impact on the risk, but it is relatively small compared with other hazardous activities, like smoking tobacco.

How these principles of basic physics were used to develop powerful weapons is what I turn to next.

Chapter 2

FROM BASIC PHYSICS
TO AWESOME WEAPONS

Since the physics Frisch and Pieirls drew on was in the open scientific literature, they realised that German scientists could come to the same conclusion as they had. Horrified by the prospect of Adolf Hitler's Nazi regime developing a new and powerful weapon, they were urging the Allied forces, essentially the United Kingdom and the United States, to invest significant resources into getting the bomb first.[1] The code name for British involvement in the project was Tube Alloys. It is extraordinary, even given the restrictions on communication that were justified by wartime, that a massive research project and then a huge industrial process were undertaken in secret, with most elected politicians on both sides of the Atlantic completely unaware of what was happening. The United States had been reluctant to get involved in the war in Europe, but was forced into it by the attack from Germany's ally, Japan, on the US naval base at Pearl Harbor in Hawaii.

As discussed, uranium as found in mineral deposits is mostly the heavier and more stable isotope, uranium-238. There are small amounts of the much more radioactive isotope, uranium-235, typically about 0.7 per cent. Frisch and Pieirls concluded that only a few kilograms of uranium-235 would be needed to produce a chain reaction: the decay of one atom triggering decay in others, in turn affecting others and so on. They realised that the decay products would have slightly less mass than the uranium atoms and used Einstein's equation to calculate that the result would be a huge amount of energy released very rapidly: an explosion. Where the half-life of uranium-238 is 4.5 billion years, uranium-235 is about 700 million years. The basic idea of the proposed atomic bomb was to produce large enough quantities of uranium-235 to have a critical mass that could explode.

The technique developed in the 1940s involved combining uranium with the gas fluorine to produce a gaseous compound, uranium hexa-fluoride, and then use the small difference in mass to separate the molecules containing the heavier isotope from those containing the lighter one. The process became known as enrichment. You may have heard discussion in the media concerning the enrichment program in Iran. It has aroused political interest because nuclear reactors used to generate electricity typically use uranium in which the proportion of uranium-235 has been increased to somewhere in the range from 3 to 5 per cent. With that increase in the proportion of the more radioactive isotopes, a nuclear reactor produces enough energy from the radioactive decay processes to boil water and spin turbines to produce electricity, without the risk of an explosion. Iran has proposed enriching uranium to 20 per cent uranium-235, far beyond the level normally used for nuclear power reactors, leading to the strong suspicion that it is considering developing nuclear weapons. Neighbouring countries are understandably nervous.

At the same time as one group was developing the process to en-rich uranium on an industrial scale, a different approach was being used by other researchers. They bombarded uranium atoms with protons to produce new artificial elements, heavier and less stable. Since uranium had been named after the planet Uranus, the scientists named the new unstable elements they created Neptunium and Plutonium, after Neptune and Pluto, then known as the two outer planets of the solar system beyond Uranus. Pluto has since been downgraded and is regarded a dwarf planet, but its legacy remains in the name of an element that is intensely radioactive. It has since been discovered that very small amounts of plutonium occur naturally in uranium deposits. The most common isotope of plutonium, with atomic mass 239, has a half-life of about 25,000 years, meaning it is much more radioactive than uranium.

By early 1945, enough plutonium and uranium-235 had been produced for prototype weapons to be manufactured. The Trinity test, the world's first nuclear explosion, was triggered in July 1945 by a British scientist, Dr Ernest Titterton, whose doctoral research had been supervised by Mark Oliphant. Both those scientists were later centrally involved in the debates about using nuclear science in Australia. Oliphant returned to Australia after World War II to head the physics department at the newly established Australian National University (ANU) in Canberra. Titterton later joined the ANU as the foundation professor of nuclear physics.

The Trinity test was a plutonium bomb. Although they had done the calculation based on Einstein's formula, the scientists involved were reputedly startled by the powerful explosion it produced. Just a few weeks later, the first nuclear weapons were used by the US Air Force. A bomb using uranium-235 was dropped on the Japanese city of Hiroshima on 6 August 1945, levelling most of its buildings and killing about 100,000 people immediately. Similar numbers died subsequently from the burns and radiation they received when the bomb dropped. A plutonium bomb was then dropped three days later on the city of Nagasaki, also producing widespread damage and a very large number of civilian casualties, usually estimated at about 70,000. The devastation caused by those two weapons precipitated a surrender by Japan and the end of World War II.

Although British and American scientists had been enthusiastically involved in the Manhattan Project from the outset, there was serious division when it became clear that Germany had been defeated and the United States was proposing to use nuclear weapons against Japan. A significant group of the scientists believed that the power of the new weapons should have been demonstrated by exploding them on unoccupied islands, rather than Japanese cities. They signed what became known as the Franck Report in June 1945, forecasting that the military use of the new weapons would lead to an arms race.[2] This was a foretaste of debates that still occur today about the morality of using nuclear weapons against civilian targets.

Until World War II, most casualties of war had been the military forces of opposing sides. During World War II, German aircraft bombed British cities, not just London as the capital but other cities seen as militarily significant, such as Coventry. Many civilians were killed and injured by these bombing raids. The Royal Air Force retaliated with bombing raids on German cities, reducing much of Berlin to rubble. The firebombing of Dresden destroyed much of that city and caused very large numbers of civilian casualties. Those who supported the attacks on Japanese cities saw that simply as an extension of the pattern that had developed in Europe. Those scientists who opposed the military use of the new weapons went on to establish a journal, the *Bulletin of the Atomic Scientists*, which remains influential in discussions about nuclear weapons.

The atomic bombs dropped on Japanese cities were much more powerful than any conventional weapons. It was calculated that the explosions had been equivalent to detonating about 25,000 tonnes of TNT, then the conventional explosive used in bombs. But it soon transpired that these sorts of weapons had the potential to develop a whole new class of explosive devices, even more powerful than the bombs used in 1945 against Japanese cities. To explain the basis of these weapons, we need a little more nuclear science.

The first atomic bombs were based on a process called nuclear fission, the splitting of very heavy atoms into smaller fragments with the release of energy. As discussed, this occurs naturally when the uranium in rocks gradually decays. The Manhattan Project aimed to accelerate that natural process by artificially creating a critical mass of fissile materials, uranium-235 or plutonium-239. By 1945, it had been recognised that the enormous amount of energy produced by the sun was the result of a different process called nuclear fusion: not the splitting of heavy atoms but the joining together of very light ones. At the extremely high temperatures in the sun, hydrogen atoms can collide and join together to form atoms of helium, the second lightest element, releasing enormous quantities of energy that maintain the temperatures at which fusion can occur.

The scale of energy released is mind-boggling. Our sun is about 150 million kilometres away, but the energy it releases keeps the Earth at a comfortable temperature for life, and even relatively brief exposure to direct sunlight can burn our skin. The physicist Dr Edward Teller is usually credited with suggesting that the enormous amount of heat released by a fission bomb could temporarily create the conditions for a fusion reaction. The so-called hydrogen bomb was developed.[3] Basically, a fission device was used to produce the heat needed to cause hydrogen atoms to join together, producing helium and releasing huge amounts of energy. The first one detonated by the US military at Bikini atoll in the Marshall Islands was estimated to have an explosive power equivalent to 15 million tonnes of TNT, about sixty times the scale of the Hiroshima and Nagasaki bombs. While those weapons had destroyed comparatively small and compact cities, the new weapons made potentially available a different scale of destruction, potentially wiping out very large cities and rendering their surrounding areas uninhabitable for decades or centuries.

The warnings in the Franck Report were depressingly accurate. The Union of Soviet Socialist Republics (USSR), a communist nation centred on modern Russia, detonated a fission bomb in 1949 and a fusion bomb four years later, within a few months of the Bikini explosion. Although the US bombs had been developed by a cooperative program involving British and Canadian scientists, after the end of World War II the US Congress passed legislation prohibiting the sharing of nuclear technology with other countries. It should have been foreseeable that the United Kingdom, in particular, would resent this and start its independent nuclear operation. This is exactly what happened and Australia was soon involved. The development of fusion weapons predictably started an arms race, the consequences of which are an alarming threat to civilisation.

It is worth making one additional point about the fusion reaction. When I was an undergraduate student being introduced to nuclear science and plasma physics, there was speculation that it might one day be possible to harness the fusion reaction to provide virtually unlimited energy. The technical problems are huge, since producing a fusion reaction requires heating an amount of hydrogen gas to a temperature of millions of degrees, somehow confining it safely in that state and simultaneously extracting useful energy. Squillions of dollars have been expended building ever larger and more complex devices. The project long ago surpassed the capacity of any one country to fund and is now a global effort with unique levels of international cooperation. Some progress has been made, using incredibly powerful magnetic fields to confine the hydrogen gas in a vacuum chamber and then bombarding it with intense laser beams. In these very expensive experiments, it has been possible to produce for tiny amounts of time the conditions needed for fusion. The problem is that huge amounts of energy were needed to provide the magnetic fields and the laser bombardments.

While one recent procedure briefly provided more energy out than had been expended, we cannot expect the utopian dream of unlimited fusion power any time soon. The standard joke in the 1960s, recounted to us by the professor of theoretical physics, was 'Commercial fusion power is fifty years away – and probably always will be!' It has a ring of truth. More than fifty years later, there is still not even a credible conceptual design of what a fusion power reactor would look like, let alone a realistic timescale for building one.

Chapter 3

AUSTRALIA
AND THE BRITISH BOMBS

Mark Oliphant had been centrally involved in the Manhattan Project. In fact, Richard Broinowski documents Oliphant's unsuccessful attempt to interest the US Government in the project before the Pearl Harbor attack forced the United States into World War II.[1] While in Washington in 1941 he briefed Australia's representative there, Richard Casey, who in turn passed the information back to Canberra. Neither the Australian Government nor its public sector scientists had been aware of the possibility of uranium being turned into powerful weapons but the politicians acted decisively on Oliphant's advice, drafting the 1943 legislation that gave the government control over the country's known and putative uranium resources.[2]

When the Manhattan Project got going, there was some concern that the supplies of uranium might not be sufficient. The British Government had obtained uranium from the Congo, at that time a Belgian colony, but shipped their stock to the United States in 1940 when a German invasion appeared likely. Alice Cawte writes that the United Kingdom was asked to try to obtain uranium from parts of the then British Empire, so attention turned to Australia.[3] Douglas Mawson was apparently despatched to a site in a remote corner of South Australia, a journey beyond the road system that had to be completed by camel, to explore the possibility of providing uranium

36

for the bomb project. It turned out that the Mount Painter deposit was not as promising as had been hoped, so no Australian uranium was used in the weapons exploded in 1945.

However, when the US Congress passed 1946 legislation prohibiting the sharing of nuclear science with other countries, including the United Kingdom and other wartime allies, there was immediate pressure on the former colonies like Australia to help the British nuclear program. Perhaps it is not surprising that we were never told that we were helping the United Kingdom to acquire nuclear weapons; neither were the British public or most of their elected politicians. After the war the US Government started an 'Atoms for Peace' program, radically reframing nuclear technology, which had only been seen as producing fearsome weapons, by promising virtually unlimited energy.

The basic principle was sound. The nuclear reactors that had been used to produce fissile material for weapons also produced significant amounts of heat. A standard power station burned a fossil fuel like coal to produce heat that boiled water and then used the steam to turn a generator and yield electricity. The steam could equally as well be produced by the heat of the nuclear reactions. At the high point of enthusiasm about this new technology, it was claimed that nuclear reactors would produce electricity so cheaply that householders would not be charged for it! As a public relations exercise, designed to distract attention from the fact that the real purpose of its nuclear research program was to produce plutonium for weapons, the UK Government decided to use the Calder Hall nuclear reactor to generate small amounts of electricity. It was insignificant, about enough power to supply the needs of ten or twenty houses, but it was dressed up as a major achievement. With great ceremony and TV cameras rolling, the newly installed Queen Elizabeth II solemnly turned on the power supply.[4]

When Australian Prime Minister Robert Menzies formally opened the Rum Jungle uranium mine in the Northern Territory in 1953, he too gave the project a totally misleading slant.[5] He told the assembled media that the uranium from the mine 'can and will within a measurable distance bring power and light and the amenities of life to the producers and consumers and housewives of this continent'. Of course, that never happened. The uranium was destined to be used by the British and US governments to produce nuclear weapons. The US Government also negotiated the re-opening of Radium Hill in 1954, with a seven-year contract to supply the US military with uranium.

A second mine was opened soon after. An ore body was discovered near Mount Isa in north-western Queensland by a local taxi driver, who named it Mary Kathleen after his recently deceased wife.[6] A syndicate, Mary Kathleen Uranium (MKU), was established to exploit the deposit and supply about 4000 tonnes of 'yellowcake' to the UK Atomic Energy Authority (UKAEA). Yellowcake, the form in which uranium is usually exported, is a mix of uranium oxides dominated by a compound with the formula U_3O_8. The venture's majority shareholder was the international mining company Rio Tinto. Delivery of the uranium commenced in 1958. The contract was completed by 1963 and Pope reports that the mine was then mothballed.[7] When the Whitlam Government became interested in controlling Australian resources, the Australian Atomic Energy Commission (AAEC) acquired 42 per cent of the shares in MKU, with Rio Tinto retaining the majority of the shares. Cawte writes that this was a fallback position.[8] The government had hoped to attract Australian capital to restart the mine, then approved the AAEC obtaining the shares when the investors did not materialise. The mine operated again from 1974 to 1983, when it was closed and the site abandoned.

As the British Government secretly developed nuclear weapons, they faced the obvious problem of testing them. The Manhattan Project had used an uninhabited area in the Nevada desert to test its weapons, but the British Isles has no such remote unoccupied territory. The Australian Prime Minister in the early 1950s, Robert Menzies, had famously described himself as 'British to the bootstraps', so it was perhaps not surprising that he volunteered to allow bombs to be tested in remote parts of Australia, apparently without even consulting his Cabinet colleagues.[9]

The first British bomb test in October 1952 vaporised a frigate moored near the Montebello Islands, which most Australians would have difficulty locating – they are uninhabited islands in shallow water off the Pilbara coast of Western Australia. Broinowski writes that the site was chosen so British nuclear experts could 'simulate and calibrate the effects of a nuclear attack on a port in a British estuary'.[10] Later two bomb tests were conducted at Emu Field in the far north of South Australia in 1953.

Australian assistance was absolutely essential to the development of British nuclear weapons. Cawte quotes the British bomb designer, Dr Penney, as saying, 'If the Australians are not willing to let us do further trials in Australia, I do not know where we would go'.[11] While the tests went ahead at Emu Field, Cawte writes that the British were unhappy with the site because 'access was difficult, there was little water and dust storms were turbulent'.

A few years after that, another series of bomb tests was carried out, two in the Montebello Islands and a further seven at a new site, Maralinga, in the western half of South Australia, towards the border with Western Australia.

Ernest Titterton had by now become the foundation professor of nuclear physics at the Australian National University in Canberra, after playing a part in the development of British nuclear weapons, drawing on his experience in the Manhattan Project. He was also appointed by the government to its Atomic Weapons Tests Safety Committee, which was supposed to assure the government that the British tests were not a risk to Australian people. It later became clear that three groups had been significantly affected. Firstly, and most fundamentally, quite a few Indigenous people were still in the areas where bombs were tested and many of them suffered severe health problems as a result. The memory of those people being irradiated is a partial explanation for the hostility of Aboriginal groups to proposed nuclear activities. Secondly, the approach to protecting the Australian military personnel involved in the testing was quite cavalier. The film *Backs to the Blast* showed servicemen in shorts turning their backs before the bombs exploded, then spinning around rapidly to see the awesome spectacle of the mushroom cloud. Royal Australian Air Force personnel in shorts ate their sandwiches as they flew through the cloud collecting samples, then landed and saw British scientists in protective clothing collecting the debris from a safe distance. Finally, on one occasion a wind change caused radioactive debris to drift over South Australian settlements, including the city of Adelaide.[12]

The test sites were never properly cleaned up after the tests. A 1980s Royal Commission chaired by Senator Jim McClelland (known as 'Diamond Jim') exposed a scandalous legacy of plutonium debris at Maralinga, forcing an embarrassed British Government to contribute to a belated and incomplete clean-up. Cawte writes that the cost of the British bomb tests was 'horrendous – in the poisoned lives of many Aborigines, servicemen and miners; and in the derelict landscapes and radioactive wastelands at Monte Bello [sic], Emu Field, Maralinga, Radium Hill and Rum Jungle'.[13]

History will record that Australia was centrally involved in the development of British nuclear weapons. We supplied uranium and we cheerfully provided the land for the bombs to be tested. At the time, we were still part of the British Empire. I remember my primary school celebrating Empire Day on the anniversary of Queen Victoria's birthday, 24 May. Each of us was given a little card with two crossed flags, the Australian flag and the Union Jack. Of course, the Union Jack remains in the corner of the Australian flag to this day, a reminder of our colonial history.

As well as Professor Titterton, another Australian academic who became prominent in the Australian nuclear industry had been involved in the British bomb, Dr Philip Baxter. He had been a chemical engineer working for Imperial Chemical Industries (ICI) when he was asked to produce quantities of uranium hexafluoride 'for research'. As mentioned, this was to enable enrichment of uranium to produce weapons-grade material. Baxter then went to the United States and worked on the bomb project at the Oak Ridge laboratory in Tennessee. After the war ended, he was involved in designing and building the plant to separate plutonium for the British nuclear weapons program. He came to Sydney in the early 1950s to direct the newly established New South Wales Institute of Technology, then located at the Sydney Technical College in Ultimo. Under his leadership, its status was elevated to become the University of Technology, which it was when I applied to start a part-time degree in electrical engineering there. Initially the only courses offered there were science, applied sciences and engineering, but it later expanded to add commerce, then arts and medicine. The university started a move to its new campus in Kensington in 1953 and was renamed the University of New South Wales in 1958.

Baxter believed that the economic development of Australia would require many more engineers and so he set out to increase significantly the number of students graduating as professional engineers. His experience with nuclear science was recognised by the Australian Government and they appointed him as deputy chair of the AAEC when it was set up in 1953. The establishment of the AAEC reflected a common view at the time, that nuclear energy was destined to replace older technologies like coal-fired power stations and provide virtually limitless clean energy. However, the fact that the first practical applications of nuclear science had been to produce powerful weapons was never forgotten. Throughout the following decades, in Australia and overseas, discussions about uranium and its uses were inevitably complicated by the recognition that the knowledge required to generate electricity could equally well be used to develop weapons.

That link was reinforced by some of those promoting nuclear energy in Australia, especially the two men author Brian Martin called the nuclear knights, Sir Ernest Titterton and Sir Philip Baxter.[14] Both men quite openly promoted the possibility that Australia could acquire nuclear weapons for defence purposes, either from our allies like the United Kingdom and the United States or by manufacturing them ourselves.

At the time of the UK bomb tests, there was clearly an unspoken assumption that our cooperation would lead to a reciprocal willingness by the British nuclear authorities to share their secrets and assist Australia's nuclear development. In practice, the UK scientists and their government were severely restricted by agreements entered into with the United States. Since the nuclear science had been involved in developing powerful weapons, the research was recognised as having obvious military significance and so security agencies were anxious to limit access. Those agencies also had a belief, not totally unjustified, that scientists were part of an international culture that was likely to share their ideas with colleagues, even if they came from other countries. It subsequently became clear that scientists with leftish political views had passed on vital information that assisted the communist USSR to develop nuclear weapons.

Robert Milliken's detailed analysis of the British tests and 'Australia's atomic cover-up' takes its title from the bland assurance given the Australian people by Prime Minister Menzies when he said in 1953, 'No conceivable injury to life, limb or property could emerge from the test that has been made at Woomera'. Milliken wrote:

> Although Australia political leaders at the time often presented the test to the public as joint exercises, the reality was different. Britain was in charge of them and only a handful of Australian scientists were vouchsafed information which was largely limited to safety criteria for the firing of the bombs. In the case of the minor trials, the Australians were not even involved to this limited extent. At various points in the early days of the British nuclear tests, an Australian domestic atomic energy program had been mooted as a possible spinoff from any Australian knowledge gained through its peripheral involvement with the British nuclear program; by the 1980s, however, Australia remained a non-nuclear country.[15]

Milliken argues that the Menzies Government had not really hoped to gain useful scientific insights:

> Menzies simply did not care about the tests, apart from exploiting them as a show of Commonwealth solidarity with Britain's contribution to the Cold War against communism. He was content to leave Australia's representation to the safety committee of scientists which, in turn, largely identified itself with the British scientific effort rather than with an independent Australia role.[16]

Of course, the safety committee was dominated by one scientist who had been involved in the British bomb program, Ernest Titterton, so it was hardly surprising that the group identified with the British scientific effort. In any case, as Milliken notes, they were entirely dependent on the British for information about the expected yield from the bombs and likely fall-out patterns. The result was a 'trail of deception' that 'covered up the danger to Aborigines and servicemen'. Milliken quotes the final words of the submission to the Royal Commission of the two Adelaide barristers who represented Aboriginal groups, Geoff Eames and Andrew Collett:

> The public has a right to know at what point scientific knowledge ceases to be conjecture and at what point subjectivity replaces science. The public was never placed in this position of wisdom during the British nuclear test program. The scientists cannot be heard to complain if the wisdom gained through this royal commission causes the public to be harsh in its criticism of the scientists, who did so little during the program to inform them of the risks to public safety that were being taken.[17]

Once again, the issue of the objectivity of the scientists and the trustworthiness of their advice was being questioned. It is a recurring theme in the history of Australian involvement in the nuclear industry.

Chapter 4

THE AUSTRALIAN ATOMIC ENERGY COMMISSION

The AAEC, established in 1953 by the Australian Government, was the culmination of a process that had begun in 1946, after the Hiroshima and Nagasaki bombs ended World War II. The Australian Government had decided to establish a national university in Canberra and it went to great lengths, as Alice Cawte writes, to get Mark Oliphant to return and head the physics department.[1] The biography of Oliphant points out that his financial demands for research facilities meant his appointment cost as much as all the other elements of the new university put together.[2] The expensive cyclotron constructed for his research was later dubbed by critics the 'White Oliphant'. The enthusiasm to recruit Oliphant reflected the government's acceptance of his argument that nuclear energy would power the nation's future. They had allocated significant funds to the government's science agency, then known as CSIR and later reorganised to be the Commonwealth Scientific and Industrial Research Organisation (CSIRO), for research in that field. Rather optimistically, Oliphant had told the government that it would only take three years to build a large nuclear reactor that would supply fifty to 60 megawatts of electricity, enough to power 10,000 homes or more, as well as producing plutonium that could be used for nuclear weapons.[3]

It is important to note that it was recognised at the time that the technology for providing power could also be used to develop weapons. Of course, it had evolved from the military application, so the connection was very obvious. Reflecting the concern that the science was militarily and politically significant, when the government set up the CSIRO in 1949 they removed the field of nuclear science from it. There was a feeling in government that scientists were likely to be unreliable or even leftish security risks who might pass atomic secrets to the Russians. Cawte argues that the politics of the Cold War made it impossible for the United Kingdom to cooperate with Australia in nuclear science, making it less likely that Australia would build and operate nuclear power stations.[4]

The legislation establishing the AAEC gave it quite broad powers over all aspects of the nuclear industry, from exploration for and mining of uranium to training scientists and publishing information about nuclear issues.[5] Most significantly, the law allowed the commission to construct and operate equipment 'for the liberation of atomic energy' – a very interesting expression, suggesting that the scientists were setting free energy that had been confined against its will in the uranium atoms. One of its first actions was to negotiate an agreement with the British nuclear authority to construct a small research reactor at Lucas Heights, on a site in what was then bushland well outside the suburbs of Sydney. The reactor was built quite quickly and began to operate early in 1958.

When I was a physics student in the mid 1960s, Lucas Heights was still well outside the limits of Sydney's outer suburbs. I remember scientists sharing cars for what seemed a very long journey to attend research seminars there. Since then, the suburbs have spread around and beyond the reactor, so there was some opposition from the local community when that reactor reached the end of its useful life and the government proposed building a new one on the same site. In similar terms, while it was not controversial at all for the reactor to store its radioactive waste on the site in its early days, waste storage there has become an issue of concern with suburban housing quite close to the perimeter fence of the site.

When the AAEC was established, it was clear from its title that it was expected to prepare for the construction and operation of nuclear power stations in Australia. Professor Baxter had by then become the chairman of the commission. In a remarkable 1957 comment, he described Australia as 'the last big continent which the white man has to develop and populate'.[6] Baxter continued, 'It will be a difficult task, but the full use of atomic energy should make it both easier and more certain.' Baxter was enthusiastically promoting a range of possibilities, such as a nuclear power station in remote Mount Isa to provide the electricity needed to smelt the minerals from its mine.[7] He wrote a secret memorandum in 1957, *Civil and Military Use of Atomic Reactor Fuel Cycles in Australia*, in which he suggested making it a provision of the design of nuclear power stations that they would allow production of plutonium 'for military purposes'.[8] He argued that building that possibility into the design would be a prudent step, meaning that it would only take a few months for a future government to have the capacity to produce nuclear weapons.[9] Baxter later argued publicly that Australia would need nuclear weapons for its defence.

A central component of the early work at Lucas Heights was the exploration of two possible designs for power reactors: high temperature gas-cooled reactors and liquid metal-fuelled reactors. It is worth explaining the thinking of the time.

47

The heart of a nuclear reactor is what was then called an atomic pile, an assembly of sufficient fissile material like enriched uranium to produce a chain reaction and generate energy in the form of heat. To control the reaction, control rods made of materials called moderators are used. Graphite was the usual moderator in the early British reactors. To carry the heat away and generate electricity, the early reactors used water flowing through pipes. In the 1960s, British nuclear engineers suggested that it would be better in principle to use a gas to carry away the heat from the nuclear reaction. The main argument was one of safety. As water gets hotter, it becomes less able to carry away heat. This posed the worrying possibility that the reactor core over-heating might make the cooling less effective, leading to further heating and a potentially catastrophic accident. By contrast, gases become more able to carry heat as they get hotter, so the scientists argued that a gas-cooled reactor would be intrinsically safer.

When I was a young scientist working for my doctorate in England, the UK authorities were building a sequence of new power reactors called Advanced Gas-cooled Reactors, or AGRs. The AGR program turned out to be a disaster, with basic engineering errors, long delays and huge cost blowouts.[10] I remember one extreme example of a problem that occurred in the construction of the Dungeness B reactor. A concrete shell was poured to hold the steel pressure vessel in which the reactor would be housed; the point of a pressure vessel was to ensure that any accident would not release radioactive material into the surroundings. When the moment came to install the pressure vessel, it was found that it was larger than the concrete housing meant to accommodate it! There was public amazement that such a basic schoolkid error could happen in the application of advanced nuclear technology.

The case for liquid-metal reactors had a similar rational basis. In the 1950s, nuclear industry sources had lamented the fact that the average efficiency of power stations based on water-cooled reactors was only about 20 per cent. In other words, only about 20 per cent of the heat energy coming from the reactor was actually turned into electricity to be exported from a typical nuclear power station. Using a liquid metal such as sodium as the coolant, it was argued, would enable higher temperatures to be achieved, with an improvement in efficiency to somewhere around the 40 per cent figure typically achieved now by recently built coal-fired power stations. As with the gas-cooled reactors, the basic idea was sound but the engineering practicalities have proved formidable. I will discuss later the most ambitious venture in this area, the fast breeder reactor. As I mentioned earlier, my doctoral research was funded by the UKAEA group building the Prototype Fast Breeder Reactor in the far north of Scotland. That project has since been abandoned, after decades of work and the expenditure of huge amounts of public money. In 2007, the International Atomic Energy Agency (IAEA) observed that more than US$50 billion had been spent trying to develop a commercial sodium-cooled fast breeder reactor.[11] While there was then still some optimism in the IAEA about the prospects, and some nuclear enthusiasts in Australia are still lauding the possibility of a fast breeder reactor, most of those development programs overseas have now been abandoned, the engineering problems proving intractable.

In retrospect, the AAEC backed two dud horses in the race to develop practical power reactors. As those who support nuclear energy point out, forty countries now have between them more than four hundred nuclear power stations. Nearly all are so-called Light Water Reactors, which use ordinary water as the coolant. While they are less efficient than a liquid metal cooled reactor theoretically would be, in practice they have proved much easier to build and so have been less expensive. That being said, economics remains a fundamental problem for nuclear power.

Because the AAEC was set up in the era of Cold War politics and in the atmosphere of nuclear technology being seen as of obvious military importance, there was always a security element to its operation. My predecessor as Director of the Science Policy Research Centre at Griffith University, the late Ann Moyal, published a famous study of the AAEC in 1975.[12] She was strongly critical of the way the commission had been an unchallenged source of advice to government on nuclear matters, with very limited capacity in government to oversee its work and ensure that it was operating in the public interest. Broinowski writes of the concern in some government departments about the fact that Professor Baxter was really the sole source of advice on nuclear issues.[13] It became a subject of considerable anxiety when there was discussion within government in the early 1970s about the United Nations proposal for a treaty to prevent proliferation of nuclear weapons.

During the 1960s, as chairman of the AAEC, Professor Baxter was a prominent public advocate of the case for developing nuclear power in Australia. In 1969, for example, he confidently predicted that Australia would have 44 gigawatts of installed nuclear power by the year 2000.[14] To put that figure in perspective, it is a lot more than the maximum total demand in the national electricity system in 2021! When I did a retrospective analysis in 1984 of attempts to introduce nuclear power to Australia, the most obvious conclusion was that it was complicated by our federal system. When the states federated in 1901 to form a nation, the constitution ceded some specific powers to the newly formed Commonwealth: foreign affairs, defence, 'posts and telegraphs' most obviously. The founding document said specific-ally that those areas not specified would remain the province of the states. At the time of the establishment of the AAEC, there was no such thing as a national electricity grid – even today, the so-called 'national grid' does not extend to Western Australia or the Northern Territory. In the 1960s, each state managed its own electricity system. There were even subsidiary authorities. Brisbane City Council ran two power stations at Bulimba and Tennyson to provide the city's electricity supply until the Queensland Government intervened, largely to overturn what it regarded as the unfair situation that power was cheaper in Brisbane than in country areas. As a relatively compact city in the 1970s, Brisbane had an electricity system with relatively low distribution costs, whereas rural areas had both fewer people and longer supply lines.

The federal system provided two serious obstacles to the building of nuclear power stations. First, there was no prospect even in prin-ciple of a national program. Each state was making its own decisions based on its resources, most obviously black coal in Queensland and New South Wales, and brown coal in Victoria and South Australia. The second problem was one of scale. For reasons of stability in a grid system, it is undesirable for any one unit to provide too large a fraction of the power. There is also an economic issue.

Since it is necessary for the system to continue to operate when there are planned or unplanned outages of any unit, the backup must be able to cope with the unavailability of the largest unit in the system. Because a nuclear reactor is a much more complicated way of heating water than burning coal, the capital cost is a fundamental obstacle unless the power station is very large. There are economies of scale; a 1000 megawatt power station does not cost ten times as much to build as a 100 megawatt installation. For this reason, nuclear power stations were competitive with coal in overseas systems such as the United Kingdom if they were very large, but the sizes of the state electricity networks in the 1960s meant that such large power stations would have been unsuitable additions. Power stations small enough to meet the criteria for the state grids could not be built and operated at costs that would have allowed them to compete with coal, even the low-grade coals being burned in Victoria and South Australia. So the AAEC and scientists who were advocating nuclear power stations were facing what would now be described as strong economic headwinds. They just could not make a case that the power would even be competitive with coal, let alone so cheap it would not be metered.

South Australia's long-serving premier, Thomas Playford, was consistently enthusiastic to have a nuclear power station in his state, at least partly because they only had poor quality coal. I remember visiting power stations in South Australia when I was a member of the National Energy Research, Development and Demonstration Council in the 1980s. I was appalled to see the amount of pollution produced by the local coal, containing very large quantities of waste materials. But the state's relatively modest electricity demand meant that nuclear power was totally uneconomic.

Long-serving Queensland Premier Joh Bjelke-Petersen was quite keen to have Commonwealth Government support for nuclear power, although he was reportedly reluctant to have the power station within his state! It has subsequently emerged that he was keen to establish other nuclear facilities in Queensland, with secret negotiations to try to procure Commonwealth support for a uranium enrichment plant near Rockhampton.[15] It would have been the largest industrial investment ever in his state. That proposal was canvassed very quietly with the short-lived Gorton Liberal Government in the late 1960s, but was shelved when the prime minister was replaced.

The AAEC did huge amounts of research into the various approaches to uranium enrichment and constantly promoted the idea that Australia should add value to uranium by enriching it, rather than exporting the metal oxide. It kept surfacing from time to time. I was at an angry rally that filled Caboolture town hall during the 1983 election campaign, when a leaked report said that the Fraser Liberal Government might fund an enrichment plant in that area just north of Brisbane if they were re-elected. The local community was worried about the impact of such an industrial venture on their main economic activity of dairy products. There was also some understandable concern about the huge gamble with taxpayers' money such a venture would have been. I recalled a Victorian physicist observing, after a study of the industry, that wherever uranium was enriched, the local taxpayers were impoverished. The huge subsidies had historically been justified because enriched uranium was used for nuclear weapons. Politicians generally tend to believe that defence justifies almost unlimited public funds, so no expense should be spared if the military can argue that the defence of the country is at stake.

It is worth a digression explaining the technology of uranium enrichment. The basic idea was developed as part of the Manhattan Project to provide the enriched uranium for the first nuclear weapons. As discussed earlier, naturally occurring uranium is mainly the isotope uranium-238, with smaller amounts of the more active isotope uranium-235. The process of enrichment works to increase the fraction of uranium-235 from the normal 0.7 per cent to between 3 and 5 per cent for the uranium to be used in nuclear reactors, or to a much higher percentage for weapons. That is why enrichment technology is politically sensitive and usually subject to security restrictions, as the same technology that is used to enrich uranium for use in reactors can be extended to make weapons-grade material.

The two common techniques are diffusion or centrifuge. Both use the slight difference in mass between the two isotopes. Uranium is reacted with fluorine to form a gas, uranium hexafluoride, UF_6. In the diffusion technique, the gas is passed through a porous membrane and the slight difference in mass means the lighter molecules move marginally more rapidly through, so the gas on the output side is slightly enriched. That process is repeated thousands of times to produce the desired level of uranium-235. In the alternative and now more common process, the gas is introduced into a rapidly rotating centrifuge and the molecules containing slightly heavier uranium-238 are flung to the outside. As Pope points out, it is exactly the same process used by dairies to separate cream from milk.[16] I can remember using that sort of separator after milking cows on my grandparents' farm, turning the handle vigorously and seeing the lighter cream emerge from one spout and the heavier milk from the other. As with diffusion, the process has to be repeated very many times because the difference in mass is quite small, less than 1 per cent.

So uranium enrichment is an expensive process because it uses huge amounts of electricity. The AAEC was eventually told to discontinue the research and development effort when Bob Hawke's Labor Government was elected in 1983, implementing that part of the ALP policy that opposed involvement in the nuclear fuel cycle beyond mining, milling and exporting uranium in the form of yellowcake.[17]

The AAEC had not been encouraged to pursue the option of nuclear weapons by Robert Menzies as Prime Minister. In 1957 he had expressed the view that it would be good for the world if nuclear weapons were only in the hands of the United States, the United Kingdom and the then USSR, on the interesting grounds that those nations were so well informed about the power of those bombs that they would be reluctant to use them. Broinowski observes that momentum for Australia to develop nuclear weapons built up during the 1960s, with China conducting its first weapons test in 1964 and the United Kingdom deciding to withdraw from the region east of Suez in 1967.[18] Various Liberal Party backbenchers had been keen for Australia to be nuclear-armed. As early as 1960, W.C. Wentworth had argued that we could not rely on the United States to defend us, so we needed our own nuclear weapons.

In 1965, Professor Baxter was despairing about state governments taking the plunge on nuclear power, so he briefed Canberra public servants about alternative nuclear reactors, emphasising their capacity to produce plutonium as well as electricity. He talked about three generations of reactors, of which the Advanced Gas-cooled Reactor was seen at the time as the most sophisticated. Broinowski says 'He couched his language in terms of electricity generation, but his subtext was the production of weapons-grade plutonium'.[19] The Minister for National Development, David Fairbairn, subsequently took to Cabinet a proposal for a 250-megawatt reactor as the first step towards the department's goal of seven or eight reactors in New South Wales and Victoria. He noted that the design he supported would produce 'considerable quantities of plutonium', but the Menzies era was nearing its end and the project did not go ahead at that stage.[20]

Later in the same year, a group of backbenchers, including my local member, Jeff Bate, urged the Cabinet to authorise production of nuclear weapons.[21] Bate later achieved a modicum of fame when he married

Zara Holt some time after her husband, Prime Minister Harold Holt, disappeared when swimming in rough surf conditions at Cheviot Beach in 1967. The loss of Holt led to the Coalition Government choosing as its leader and replacement prime minister John Gorton, who was quite enthusiastic about both nuclear power and the prospect of Australia developing nuclear weapons. Gorton approved a proposal by Professor Baxter for the construction of a nuclear reactor on Commonwealth land at Jervis Bay, on the New South Wales coast. Support for the possibility of an Australian bomb had increased after the 1968 Tet offensive in Vietnam, when perceptive observers realised that the US attempt to impose a puppet government on Vietnam had failed. There was an increasingly widespread view that Australia had to manage its own defence without assuming that the United States would defend us against any threat.

Despite the covert agenda to give Australia the capacity to develop nuclear weapons, the Jervis Bay project was always problematic. The government went ahead and called for tenders for construction of the reactor, but there was significant opposition in Canberra. Treasury did some calculations and estimated that the electricity from the reactor would be about twice as expensive as that from operating coal-fired power stations, so they could not see the state electricity bodies wanting the high-priced power and cautioned against the project.[22]

After three turbulent years as Prime Minister, Gorton was removed in the sort of internal coup that has occurred with increasing frequency in recent years. His replacement, William McMahon, had briefly been Minister for Foreign Affairs. In that capacity, he had been persuaded that Australia should sign the Nuclear Non-Proliferation Treaty (NPT) that was being developed by the United Nations. For that reason, he did not support clandestine production of material for nuclear weapons and ordered a review of the Jervis Bay project. That review predictably found a cost blow out, with every one of the tenders submitted showing a significantly higher capital cost than had been budgeted. The McMahon Government suspended the proposed reactor.[23] Within a year, the Whitlam Government had been elected and the proposal was scrapped, but the irrepressible Professor Baxter was still saying confidently that Australia would begin building nuclear power stations within the next ten years. He also continued to campaign for nuclear weapons. In 1976 he said, 'We need a varied array of missiles of short and medium range, some of which should carry nuclear warheads'.[24] Professor Titterton also argued that we should consider 'either by acquiring from our allies or making for ourselves, a nuclear weapons stockpile sufficient for us to be able to repel any attempted invasion'.[25]

The AAEC was not just evaluating possible nuclear power reactors. It was also considering what would now be regarded as ridiculous proposals to use nuclear devices for large-scale civilian construction projects. The US nuclear authority had developed a program called Operation Plowshare, the idea of which was to use nuclear explosions to move huge volumes of earth and produce artificial harbours. The AAEC was interested in this idea. A Canberra task force, chaired by Professor Oliphant, was set up in 1961 and negotiations began with the Americans. Broinowski documents that a serious plan had been developed by January 1969 to use US nuclear explosives to create a huge new harbour on the north-west coast of Western Australia at a site called Cape Keraudren.[26] The idea was that five bombs would be detonated, each about ten times as powerful as the Hiroshima bomb, to produce an artificial harbour 2.4 kilometres long, 200 metres wide and 35 metres deep. Broinowski notes that the level of concern at the time was mild, 'given the horrendous nature and scale of the prospective explosions and the possibility that radioactivity might contaminate the food chain'.[27] There was some international concern about whether such a project was consistent with the draft non-proliferation treaty, despite the argument that this was a totally peaceful project, since it would have required developing bombs. The harbour would have enabled a US mining company, Sentinel, to export iron ore, but the company was not proposing to pay any of the costs and wanted assurances that it would have a radiation-free loading port.

A few months later, the relevant Canberra minister issued a statement saying the project had been cancelled, as Sentinel no longer wanted to go ahead. Broinowski reports that the AAEC and its US counterpart continued to explore other options, including harbours at various sites: Limestone in the Great Australian Bight, three possible sites in Tasmania and another on the WA coast. There was even, he writes, a 'bizarre proposal … to use a hydrogen bomb to shatter a rich vein of iron ore at Wittenoom Gorge 700 metres from a blue asbestos crushing plant'.[28] He concludes that 'most Australians with some knowledge of nuclear physics were thankful that none of these, especially the last one, went beyond the stage of a feasibility study'.

During the 1970s, the economics moved decisively against nuclear power. Improvements in both the mining of coal and the operation of coal-fired power stations had reduced the cost of providing electricity from coal. At the same time, the nuclear power industry in the United Kingdom had been plagued by cost overruns. There has been no serious proposal to consider nuclear power in Australia since the cancellation of the Jervis Bay project in 1972. By the time the Australian Science, Technology and Engineering Council (ASTEC) conducted a review of Nuclear Science and Technology in Australia, it was clear that the main function of the Lucas Heights reactor was to produce radioactive materials for medical and industrial use. That ASTEC review supported a proposal that the 1953 Atomic Energy Act, which established the AAEC, should be repealed and replaced by legislation to set up a body to be called the Australian Nuclear Science and Technology Organisation (ANSTO).[29] That subsequently took place, formalising the three roles of ANSTO: to pursue nuclear science and technology research and development, to provide relevant products and services, and to advise government on nuclear matters.

There was vigorous debate when the original Lucas Heights reactor was nearing the end of its useful life. The argument for a replacement reactor did not mention either the possibility of building nuclear power stations or the manufacture of nuclear weapons.[30] The main emphasis was on the need to produce nuclear materials for medical purposes, both diagnostic imaging and the treatment of cancers. A secondary argument was the benefit of maintaining an ability to do research in nuclear physics. For all practical purposes, the debate about whether Australia should have nuclear power stations ended fifty years ago, although a few zealots have never stopped urging governments to 'go nuclear'.

Chapter 5

RANGER, THE FOX REPORT AND URANIUM EXPORTS

On my desk there is a well-thumbed copy of the Fox Report, the 1976 volume that is formally entitled Ranger Uranium Environmental Inquiry First Report. It is a constant reminder of the new turn in the public debate about nuclear issues. Relatively small uranium mines at Rum Jungle in the Northern Territory and Mary Kathleen in Queensland had been quietly producing relatively small amounts of the metal for overseas sale, mainly to the United States and the United Kingdom for nuclear power stations or weapons production. Australia had signed the UN Nuclear Non-Proliferation Treaty, so in principle we could only sell uranium to the five countries that already had nuclear weapons – the United States, the United Kingdom, the then Soviet Union based on modern Russia, China and France – or to other countries purely 'for peaceful purposes'. The inquiry resulted from the discovery of a large ore body at Ranger in the Northern Territory, near both the Arnhem Land Aboriginal Reserve and the proposed Kakadu National Park.

The AAEC and a new company, Ranger Uranium Mines Pty Ltd, developed a proposal to mine and mill the uranium for export to countries operating nuclear power reactors. A memorandum of understanding between the AAEC and two companies, Peko Mines Ltd and Electrolytic Zinc Company of Australasia Ltd, formalised their relationship as partners in the new company, with the AAEC to pay 72.5 per cent of the cost of the project and the two commercial partners equal shares of the remainder.[1] I find it interesting, in retrospect, that the Australian Government thought this project was sufficiently germane to the national interest to support public funds paying most of the cost. In July 1975, the Whitlam Government set up a commission to inquire into environmental aspects of the proposal. Justice Russell Fox of the ACT Supreme Court was appointed to preside over the inquiry, with two other commissioners: Graham Kelleher, a distinguished civil engineer who is now a Fellow of the Australian Academy of Technology and Engineering, and Charles Kerr, Professor of Preventive and Social Medicine at the University of Sydney.

As the first report says in its 'Preface', many of the submissions to the inquiry raised issues much broader than the direct environmental impacts of the proposed operation, including the possibility of accidental nuclear explosions, the risk of fissile material being obtained by terrorists and the problem of managing radioactive waste from nuclear reactors. The commissioners decided to tackle those issues in a first report, with the subsequent report dealing with the specific environmental aspects of mining and milling uranium at Ranger.

It is worth commenting on the scale of the proposed operation at Ranger. As the Fox Report sets out, the plan was to mine the ore and extract the uranium in the form of yellowcake, a mixture of uranium oxides with most of it in the form of U_3O_8.[2] The preliminary studies established that the ore body was about 0.25 per cent uranium oxide, so about 400 tonnes of ore would have to be mined and milled to produce a tonne of yellowcake. The proposal was for an initial production of

3000 tonnes of the final product, meaning 1.2 million tonnes of ore would have to be mined and processed each year, or about 3500 tonnes a day if the mine operated every day of the year. It is no overstatement to say that that is a large-scale mining and milling operation. The company estimated the ore body was big enough to allow that level of production to continue for twenty to thirty years, potentially starting about 1980. The inquiry was told that the overall operation would affect about 900 hectares of land, with a waste dump and a tailings dam each covering more than 100 hectares.

The report also noted that two other uranium ore bodies had been identified near Ranger, Koongarra to the south and Jabiluka to the north. At the time, there was great optimism in the mining industry for the development of these two deposits. Neither has actually gone ahead, both stymied by determined opposition from the relevant traditional owners of the land.

As discussed earlier, the Fox Report commented on the lack of objectivity of people making submissions, both in favour of the proposed operation and against it.[3] They seemed particularly surprised that 'scientists, engineers and administrators involved in the business of producing nuclear energy have at times painted excessively optimistic pictures of the safety and performance, projected or past, of various aspects of nuclear production'. The report also made the interesting observation that 'a few of the government officials who appeared before us showed reluctance in communicating matters of importance' and suggested, rather pointedly, that those officials might not have clearly understood the objectives of the relatively new Commonwealth law concerning environmental impact assessment.[4] As a final over-arching comment, the report noted that some of the supporters of nuclear development had disparaged the motivation of opponents.[5]

I later became aware of some extreme examples of this sort of argument. Professor Baxter, while head of the AAEC, dismissed opponents of nuclear energy in Australia as 'a small, well-funded, vocal minority' who used 'a mixture of untrue and hysterical statements, emotionally concocted to frighten the lay public'.[6] He later went even further out on that bizarre limb when he claimed, 'The Australian anti-nuclear conspiracy is a political thing with links to international communism and the general motive of reducing the economic and military strength of the West'.[7] There was, of course, some politics in the debate about both nuclear energy and uranium exports, but those I knew who were campaigning on the proverbial smell of an oily rag would have been startled to learn that they were well funded. There was also no evidence they were part of an international communist conspiracy being directed from Moscow or Beijing. The Fox Report said explicitly that the opposition they found came 'from a wide cross-section of the general community'.[8] They also commented that they could not conclude 'that their motives and methods are any less worthy or proper, or intelligently conceived, than in general are those of supporters of nuclear development'. Their critical observation was that the subject 'is very apt to arouse strong emotions', which can lead to 'wildly exaggerated statements ... about the risks and dangers of nuclear energy' from opponents of nuclear power and 'excessively optimistic pictures' from proponents. That warning still has a ring of truth 45 years later.

The Fox Report found that the proposed expansion of uranium exports raised two important issues: the potential for fissile material to be used to produce nuclear weapons and the need to manage the radioactive waste from reactors. 'The nuclear power industry is unintentionally contributing to the risk of nuclear war,' it said, recommending that uranium exports should be strictly controlled to prevent weapons proliferation.[9] It also noted that the 1976 Flowers Report from the UK Royal Commission on Environmental Pollution had argued that development of nuclear power should be limited until it had been demonstrated that radioactive waste could be 'safely contained for the indefinite future'.

The Fox Report sparked vigorous debate in Australia, with community groups sponsoring public discussions, taking note of the report's conclusion that their 'findings of fact' could be more valuable than their recommendations. The report argued that the main purpose of the inquiry was to provide the public with reliable information 'so that they can form their own opinions'. And they did.

I remember being on a panel one steamy Friday night in February 1977 in the town hall at Nambour, a small town in the Sunshine Coast hinterland, where a public meeting had been convened by their Apex Club. They approached me and three other speakers from different backgrounds in the spirit of helping the local community to appreciate the issues and form their own opinions. About two hundred people turned up to witness a discussion that became quite heated. The general mood of the meeting was initially clearly in favour of the proposal to mine and export uranium, with its promise of jobs and export income. But the speakers from the Uranium Producers Forum over-egged the pudding and made quite indefensible assertions about the economics of nuclear power, the management of waste and the security of fissile material. I was vigorously attacked by the uranium mining company representatives when I quoted from the UK nuclear industry house journal *Atom* to show that they were lying to the meeting.

The Fox Report led to serious division within the ALP. In a precursor of the contemporary differences about the proposed Adani coal mine in northern Queensland, those on the left of the party mostly opposed the mining and export of uranium on moral grounds, while those on the right generally supported the potential jobs that would be created. In 1977, after a heated debate and frenzied lobbying on both sides, the ALP national conference adopted a policy opposing expansion of uranium mining.[10] But the Whitlam Government, which began the inquiry, had been removed from office in 1975. Under Malcolm Fraser, the Coalition Government was enthusiastic to see the Ranger mine go ahead and actively encouraged other possible export ventures. Fraser tried to elevate the program to a moral issue, claiming 'an energy-starved world' needed our uranium.[11] He also stated that the waste problem had been solved.[12] That was a barefaced lie.

Since it was not prudent for a young scientist to accuse the prime minister of lying, I described it instead as 'a very modest announcement of a great scientific advance'. Of course, the problem had not been solved; over forty years later, it is still an issue. Huge volumes of nuclear waste are stockpiled at power stations around the world. Sweden and Finland have adopted a good process of community involvement and are well on the way to a potential solution involving storage in deep underground repositories – but the issue remains contentious everywhere else.[13] The UKAEA has made several attempts to enlist community support for possible sites to establish a repository, without success at the time of writing. For decades, the US nuclear industry had invested in a site with the delightfully appropriate name of Yucca Mountain, but the Nevada administration eventually turned against it. One Australian veteran anti-nuclear campaigner observed that even the state that houses Las Vegas, and is critically dependent economically on that casino industry, was not prepared to gamble on the safe storage of nuclear waste.

At the time of the Fox Report, the nuclear power industry was looking strong.[14] At the end of 1975 there had been 157 power reactors operating in nineteen different countries. The total installed electrical capacity of nuclear power stations at the time was about 72 gigawatts – to put that figure in perspective, it is about twice the 2021 capacity of the Australian national grid. Another thirteen countries had nuclear power stations under construction or on order. I was working in the United Kingdom at the time and there was a vigorous debate going on about the electricity industry and its forward planning. There were two significant issues: the scale of capacity needed and the competition between coal and nuclear. The country already had twenty-eight operating nuclear power stations and another eleven under construction, with a clear trend towards increasing the size of individual installations – the eleven being built were designed to produce more power than the twenty-eight already operating. Some in the electricity industry were arguing for a program of another thirty-six nuclear power stations to meet the projected future demand.

The first big issue was the demand for electricity. The UK Open University's Energy Research Group, of which I was an active member, did research that showed future demand was much less than was being projected by the industry body, the Central Electricity Generation Board (CEGB).[15] Total demand in the United Kingdom had increased steeply in the 1950s and early 1960s, as households electrified and installed new appliances like electric cookers and refrigerators. The CEGB had extrapolated this trend and was anticipating a continuation of the rapid growth in demand. By the early 1970s, it was clear that the growth in demand had slowed. Once every household had their stock of electric appliances, there was only slow increase driven mainly by increasing population. The electricity system actually had surplus capacity and there was no need for new power stations. The planned program of thirty-six more power reactors would have required massive amounts of energy for construction, so it would have created an energy shortage.[16] The planners would then have been able to say that they were meeting the demand they had actually produced!

The second issue was whether coal or nuclear would give cheaper electricity. As I pored over reports prepared for the UK Government, I was puzzled by an obvious discrepancy. Those promoting nuclear energy had published peer-reviewed papers showing it was the cheaper approach, but those who opposed it had also published peer-reviewed papers showing it was more expensive! How could that be? I wondered if some of the people involved were just making it up, using creative accounting to produce the answer they wanted. It wasn't that crude, although some vested interests were clearly working to get the answer that suited them. The problem is that there is no agreed basis for making an honest comparison over the lifetimes of power stations using different technologies. It costs more to build a nuclear power station than its coal-fired equivalent because a nuclear reactor is a more complicated way of boiling water than burning coal. But the running costs are much higher for a coal-fired power station, which requires the mining and transport of millions of tonnes of fossil fuel. Any comparison needed to make estimates of future coal prices to determine whether they offset the higher capital costs of nuclear power – and different credible estimates of future prices lead to different estimates of overall costs.

There is also a more fundamental problem involved in calculating overall lifetime costs. The comparison requires choosing what economists call the discount rate, recognising that future costs and revenues have a reduced value. The whole money-lending industry reflects our preference for funds available now over funds available at some time in the future. We are prepared to borrow $10,000 to have it now, even if that means paying back $11,000 in a year's time. There are different views about appropriate discount rates for future costs, and some analysts clearly had a vested interest: those who worked for the coal industry or in nuclear organisations understandably chose assumptions that gave their preferred overall results. It remains difficult to determine a legitimate basis for comparing the long-term costs of different energy supply technologies, because we cannot know what the operating costs will be in thirty years' time.

There was a third complication that was not recognised at the time. Whether the technology being considered is nuclear or coal, there are economies of scale. Doubling the size of the power station does not double the capital cost. It still requires one site, one construction activity, one facility for receiving and processing fuel, one connection to the electricity grid and so on. It was clear that a 2-gigawatt power station would deliver cheaper electricity than a smaller installation, so the CEGB embarked on a systematic program of increasing the scale of power stations, assuring the public and political decision-makers that that would drive power prices down. In practice, it didn't. What was ignored in the planning was the need for the system to be able to cope with the loss of any unit. A power system has what is called spinning reserve, basically extra generating capacity standing by to cope with an increase in demand or the loss of a supply unit. As the scale of individual power stations increased, it became necessary to have more capacity standing by in case of a failure. When an overall analysis was conducted, it found that the expected lowering of prices did not eventuate. While the cost of generating each unit of electricity had been reduced by the economies of scale in larger power stations, these savings had been cancelled out by the increasing cost of running the overall system to cater for any unplanned outages.[17]

At the time the Fox Report was written, electricity authorities were still confidently expecting power demand to keep growing by rates like 5 per cent a year, which would have led to a doubling by 1990. The case for mining and exporting uranium was based on an expectation that this growth would occur and that nuclear power would obtain an increasing share of the market, leading to significant growth in the demand for uranium. The report did caution that 'growing recognition of the overall social and economic implications of continuing growth in energy use is likely, and therefore many countries may be forced to take account of them in their energy policies'. This conclusion was a long way ahead of the thinking of governments and industry bodies that were still confidently expecting that energy use would continue to grow rapidly. A graph in the report gives various projections by the global body, the International Atomic Energy Agency (IAEA), of likely future nuclear power stations.[18]

At the time of writing the installed nuclear capacity was about 75 gigawatts (GW). In 1970, it had been projected that the 1985 capacity would be about 600 GW. By 1976, when the Fox Report was written, this had been scaled back to about 450 GW. In 1975, the IAEA had projected that the installed capacity in the year 2000 would be an amazing 2500 GW. By 1976, this had been trimmed to between 1700 and 2000 GW. Nevertheless, the lowest figure was more than twenty times the 1975 capacity, so the inquiry was encouraged to believe there would be spectacular growth in demand for uranium, justifying a large new mine.

Of course, the IAEA was not in any sense an impartial observer of the nuclear industry but more like a cheerleader. The growth has not been anything like the ridiculously rosy picture presented to the commission in 1976. In fact, the 2020 installed nuclear capacity was about 400 GW, less than the projected 1985 capacity and only about 20 per cent of the amount the IAEA had expected by 2000. The Fox Report did note that new power capacity had been commissioned when demand was increasing more slowly than had been expected and suggested that the figures could be revised downward. It also noted that small-scale generating installations 'may prove to be more economic than large coal-fired or nuclear generating units', but that 'small-scale nuclear plants are unlikely to be economic'. In those conclusions, the report was essentially pouring a modest dose of cold water on the grandiose projections of the nuclear industry.[19] The politicians in office at the time did not appear to notice this caution, which has been justified by consequent events.

The argument that was widespread in the 1970s was put most crudely in a book by the distinguished astrophysicist Professor Fred Hoyle.[20] He had made important contributions to the understanding of stars and the nuclear fusion process that gave them huge amounts of energy. He later became almost a figure of fun when he was on the wrong side of the debate about cosmology. In the 1960s there had been vigorous intellectual debate between two alternative models for the universe, known popularly as the big bang theory and the steady-state model. The first theory argued that the universe was continually expanding from a cataclysmic event about 13 billion years ago, while the second posited that new matter was being continually formed. I still remember the exciting year when two leading astrophysicists, George Gamow and Thomas Gold, came to Australia for a physics summer school at the University of Sydney and argued for the two competing models. It was the first time I had seen disagreement between scientists.

The impression I had from school was that science was a stable body of uncontested knowledge that a student had to memorise. The insoluble chlorides were lead, mercurous and silver – this had been established for decades, learn the facts. What the debates revealed to me was the reality of science, not as a fixed body of eternal truth but as an exciting exercise of trying to understand the real world, advancing theories and testing them against experiments or observations. The evidence gradually strengthened for the big bang model and the debate effectively ended when it became possible to detect the vestigial radiation at 4.2 degrees Kelvin from the big event, but Hoyle refused to accept the evidence. He found more and more fudges to try to salvage the theory he supported.

In *Energy or Extinction*, Hoyle argued that energy was the key to a civilised life, as it certainly is. We live much more comfortably than our grandparents did because we use energy either to do things that previously required muscle power, from washing clothes to digging holes or mowing lawns, or that were previously impossible, like flying overseas or moving huge amounts of information from place to place. Hoyle looked forward to a future in which billions of humans all used energy at the levels then seen only in affluent countries. He calculated that the known fossil fuel resources were inadequate to support that level of energy use, and so he concluded that there would have to be a dramatic expansion of nuclear power. I remember doing a quick calculation on the assumption that each reactor would last thirty to forty years and finding that Hoyle's hypothetical future world would need to be opening several new large nuclear power stations every day, just to replace those coming to the end of their lives.

Less grandiose, but on the same wavelength, were contributions like the 1989 booklet by Ian Hore-Lacy and Ron Hubery, *Nuclear electricity – An Australian Perspective*, which said on its cover that it was 'Issued in the interests of education by the Australian Mining Industry Council'.[21] I remember saying at the time that it was wildly improbable for the mining industry lobby group to be interested in education, and much more likely that they were interested in the profits of the mining industry. The basic argument of the book was that we needed to keep using ever-increasing amounts of energy and so we would need not just more coal but more nuclear power as well. Solar energy was dismissed as 'too diffuse and too intermittent', while wind power was seen as only practical for those very few areas with strong winds, even there requiring other sources to cope with calmer periods. The overall conclusion for these and other renewable energy technologies was: 'They therefore cannot be applied as economic substitutes for coal or nuclear power, however important they may become in particular areas with favourable conditions.'[22] Ironically, the booklet went on to suggest that there would be 'environmental objections' to large-scale use of the alternative energy technologies, while implicitly assuming there would be no such concerns about large-scale coal-fired or nuclear power stations.

As with any mineral, the selling price of uranium is affected by both the demand for the product and the available resources. The Fox Report was given estimates of the Australian resources of uranium as about 30 per cent of the world figure, with only the United States having a larger amount. Demand was obviously going to be determined by the scale of building nuclear power stations. The 1976 price for new contracts to supply uranium was up to US$40 per pound, or about US$80 per kilogram. Industry sources suggested that the projected growth in nuclear power around the world would boost demand and possibly increase the price electricity utilities would be prepared to pay, given that the price of fuel was a comparatively small component of the cost of running a nuclear power reactor. In practice, the expected rise in prices has never materialised. In 2020, the world market price for uranium was about US$24 a pound, or about US$50 a kilogram. Even without taking into account the inflation that has happened since the report was written, the price is much less than it was 45 years ago. Allowing for inflation, the slump in uranium price has been dramatic, reflecting the scaling back of plans for new generating capacity with its consequent impact on demand for uranium.

Estimates of the economic benefit of the mine relied on projections of both the scale of production and the selling price. At the time of the report, it was claimed that the mine would produce at least 3000 tonnes of yellowcake a year, with a possible expansion to double this amount. Projections of revenue and employment were given for these two cases. Table 11 of the Fox Report assumed that exports of uranium would increase spectacularly, from 2500 tonnes in 1980 to 30,000 tonnes ten years later. The lower value quoted turned out to be a good estimate for the Ranger project. In about 40 years of operation, the mine was reported by its final owners, Energy Resources Australia, to have produced 125,000 tonnes, or an average of about 3000 tonnes a year. Mining ended in 2012 and the work of restoring the land and cleaning up the site was continuing as I wrote this book. Even at the more optimistic rates presented to the commission, while it concluded that the mine 'would probably be a highly profitable venture', it also found that its overall economic impact would be small and the employment generated modest.[23] This reflects a more general observation made by Dr Richard Denniss, chief economist at the Australia Institute. Mining is a profitable industry, but it is highly mechanised and employs comparatively few people. He told a session at the 2021 Adelaide Writers' Week that more people work for Anglicare or the McDonalds fast-food chain than work in the coal mining industry.

It is worth concluding by setting out some of the findings and recommendations of the Fox Report, while noting its caution that they should be 'read and understood in the context of the Report as a whole'.[24] Obviously, I cannot reproduce here the entire report, but I recommend seeking it out in a library and reading it for more of the detail justifying these conclusions:

- The hazards of mining and milling uranium, if those activities are properly regulated and controlled, are not such as to justify a decision not to develop Australian uranium mines.
- The hazards involved in the ordinary operations of nuclear power stations, if those operations are properly regulated and controlled, are not such as to justify a decision not to mine and sell Australian uranium.
- The nuclear power industry is unintentionally contributing to an increased risk of nuclear war. This is the most serious hazard associated with the industry …
- Any development of Australian uranium mine should be strictly regulated and controlled …
- Any decision about mining for uranium in the Northern Territory should be postponed until the Second Report of this commission is presented.
- A decision to mine and sell uranium should not be made unless the Commonwealth Government ensures that the Commonwealth can at any time, on the basis of considerations of the nature discussed in this report, immediately terminate those activities permanently, indefinitely or for a specified period.
- Policy respecting Australian uranium exports, for the time being at least, should be based on a full recognition of the hazards, dangers and problems of and associated with the production of nuclear energy,

and should therefore seek to limit or restrict expansion of that production.

- No sales of Australian uranium should take place to any country not party to the NPT [the Treaty on the Non-proliferation of Nuclear Weapons]. Export should be subject to the fullest and most effective safeguards agreements, and be supported by fully adequate back-up agreements applying to the entire civil nuclear industry in the country supplied ...

- A permanent Uranium Advisory Council, to include adequate representation of the people, should be established immediately to advise the Government, but with a duty also to report at least annually to the Parliament, with regard to the export and use of Australian uranium ...

The report went on to recommend that Australia should have a national energy policy, that there should be a 'full and energetic' research and development of alternative energy sources, that there should be a national energy conservation program, that the policy regarding uranium export should be reviewed regularly and, perhaps most fundamentally, that 'there should be ample time for public consideration of this Report, and for debate upon it'. I will reflect later on the extent to which these recommendations were carried out.

In a postscript to the report, it observed that the UK Royal Commission on Environmental Pollution had just conducted an inquiry into the environmental risks of the nuclear industry, under the direction of Sir Brian Flowers.[25] The report had only been presented to the UK Parliament in late 1976, after the Fox Report had been written, but it added the postscript for the purpose of drawing attention to the very valuable discussion which the UK Report contains, especially with regard to two important issues affecting the nuclear industry: managing radioactive wastes and preventing weapons proliferation. Four quotations from the Flowers Report are particularly important:

> There should be no commitment to a large programme of nuclear fission power until it has been demonstrated beyond reasonable doubt that a method exists to ensure the safe containment of long-lived, highly radioactive waste for the indefinite future.

> ... the spread of nuclear power will inevitably facilitate the spread of the ability to make nuclear weapons and, we fear, the construction of these weapons.

> ... we see no reason to trust in the stability of any nation of any political persuasion for centuries ahead. The proliferation problem is very serious and it will not go away by refusing to acknowledge it.

> The dangers of the creation of plutonium in large quantities in conditions of increased world unrest are genuine and serious. We should not rely for energy supply on a process that produces such a hazardous substance as plutonium unless there is no reasonable alternative.

Reading again the principal findings and recommendations of the 1976 Fox Report, as well as the postscript drawing attention to the UK Flowers Report, it is astonishing that one newspaper saw it as 'a green light for yellow-cake'. The report was at best a cautionary yellow light, sounding a loud and clear warning about the potential consequences of exporting uranium and contributing to the serious risks of weapons proliferation and waste management.

As I noted in the Introduction, we all see the world through the lenses of our values and experience. Governments and newspaper editors have their particular predispositions and, as we all do to some extent, see what they want to see. Others interpreted the report differently and a vigorous political debate ensued.

Chapter 6

THE POLITICS
OF URANIUM IN THE
1970S AND 1980S

I had been aware of nuclear power being a political issue when I was still a student. In 1964, the UK Labour Party led by Harold Wilson won a general election after a long period of government by their Conservative Party, broadly equivalent to Australia's Liberal Party. A significant part of their election campaign had been their argument that 'the white heat of the technological revolution' would bring about a modernisation of Britain and new economic opportunities.[1] They contrasted their enthusiasm for new technology, especially nuclear power, with the grouse-moor image of the Conservative Party, portraying them as out of touch with the new reality and unaware of the opportunities it provided. By the time I went to the United Kingdom to do research for my doctorate in 1968, the honeymoon was over and there were serious divisions inside the Labour government about the competing alternatives of nuclear energy and coal-fired power.

The political reality was that the coal industry employed a very large unionised workforce and their union was a strong voice within the broad labour movement. By contrast, nuclear power stations employed comparatively few people; equally importantly, the scientists and engineers working in the nuclear industry were less likely to be members of a union. So the UK trade union movement was working hard to reduce the government support for nuclear power, seeing it as a direct threat to jobs in the coal industry. There had been no parallel conflict in Australia, most obviously because no proposal for a nuclear power station had reached the point of threatening coal mining jobs.

The Fox Report catalysed a serious debate about Australia's role in the nuclear industry. There had been no significant opposition to mining and export of uranium for nuclear power stations in the United Kingdom and the United States, while the Australian public had certainly not been told that our uranium was helping produce the British bomb. Even the testing of their nuclear weapons on Australian soil had aroused relatively little comment, although I know many people in South Australia were not impressed when it became clear that contamination from one of the tests had reached Adelaide. The much more serious impacts on Indigenous people living near Emu Field and Maralinga had gone unremarked at the time, although they became a political issue in the 1980s.

The testing of nuclear weapons in the Pacific did arouse concern in Australia. While there was relatively little criticism of the US tests, there was hostility to the weapons detonated by France in the Pacific Ocean atolls they controlled. Broinowski documents the extent of atmospheric testing of nuclear weapons before it became a political issue: 'up to 1963 the United States carried out 183 atmospheric nuclear tests, the Soviet Union 118, the United Kingdom 18 and France 4'.[2] Clearly there was an element of selective indignation in Australian politicians being more concerned about the four French tests than the much larger numbers carried out in the southern hemisphere by the United States and the United Kingdom. The United Nation's 1963

Partial Test Ban Treaty prohibited testing in the atmosphere, but allowed continued bomb tests underground.

During the period in office of the Whitlam Government it became an issue because France and China were both still testing nuclear weapons above ground. Australia joined New Zealand in mounting a legal challenge to the French tests at the International Court of Justice in The Hague.[3] While France did not recognise the authority of the court to restrict its weapons testing, it did agree that future tests at Mururoa atoll would be underground. There was political criticism of the Whitlam Government for being less vigorous in their criticism of Chinese test. They argued that the Chinese tests, being in the northern hemisphere, had less impact on Australia, while also making the point that action in the International Court of Justice was not possible because China did not recognise its authority at all. Critics said that the government was more cautious about antagonising China, with its obvious importance to the region, than France as a colonial power retreating from the Pacific.[4] In both cases, government officials were apparently worried that criticism might affect Australia's trade relations. More generally, there has always been a selective element to criticism of weapons testing. As an extreme example, I can remember being involved in discussion in the 1960s of the threat to the community of nuclear weapons at the time of the huge protest marches organised in the United Kingdom by Citizens for Nuclear Disarmament (CND). I was startled to hear one person's assertion that China should be exempted from the criticism because they were developing 'a people's bomb'!

There was a level of concern within the Australian Labor Party about uranium exports, probably influencing the ALP Government to commission the Fox Report. This did not indicate hostility to the idea of exporting uranium. Broinowski reports that Whitlam as Prime Minister tried to encourage Japan to buy Australian uranium, while one of his ministers even went to Iran to try to persuade the Shah of Iran to see Australia as a source for his nuclear program.[5] Given recent concern about that country's nuclear ambitions, it is worth reflecting that the United States were strongly supporting Iran's nuclear development program before the Shah was deposed in the Islamic Revolution, while Australia was keen to sell our uranium there. It was only an accident of history that the Shah was deposed before the United States had finalised its plan to supply nuclear reactors and Australia had gone ahead to provide the uranium to fuel them.

Broinowski describes the Fox Report as 'the most significant public environmental inquiry ever undertaken in Australia'.[6] It was widely read and its cautious conclusions quoted to make a case against the proposed expansion of uranium mining and export. As an example, I have a battered copy of a booklet produced in 1977 by four trade unions, describing uranium as 'one of the most significant issues facing Australian workers and the Australian people'.[7] After noting that the issue had moral, economic and environmental dimensions, the booklet concluded that 'on each of these grounds, there is very firm justification for the Australian people to oppose uranium mining'. The core of the booklet is an essay cataloguing the risks of the nuclear industry, prepared by the environment and energy committee of the Amalgamated Metal Workers and Shipwrights Union, and a contribution from Dale Bridenbaugh, a US nuclear engineer who resigned from General Electric when he became disenchanted with the industry as 'very unsafe and very risky'. Together, these make a solid case for being cautious about exporting uranium for use in nuclear reactors.

The industry was also opposed by mainstream environmental organisations like the Australian Conservation Foundation (ACF), within which there must have been some interesting conversations, because in 1977 the distinguished scientist Sir Mark Oliphant had replaced Prince Philip, Duke of Edinburgh, as its President. Oliphant was very sympathetic to the idea of nuclear power, so he would probably have supported uranium mining and export, a stance that would have put him at odds with most of the members of ACF's governing council. A range of community groups emerged to oppose mining. There was a small group called Campaign Against Nuclear Power in Queensland. The Melbourne-based Movement Against Uranium Mining (MAUM) organised major rallies that attracted as many as 50,000 people. As noted, there was also a serious division within the ALP on the subject.

In 1977 I heard former federal ministers identified with the left of the party, such as Tom Uren and Jim Cairns, give speeches opposing uranium mining, siding with the unions that had produced the above-mentioned booklet. Other unions like the Australian Workers Union were strongly in favour of the industry and supported by equally distinguished former ministers identified with the right of the party, such as Bill Hayden and Paul Keating, as well as the head of the Australian Council of Trade Unions (ACTU), Bob Hawke. At the 1977 ALP National Conference in Perth, there was a heated debate. The context was that the Whitlam Government, which commissioned the Fox inquiry, had been controversially deposed in 1975 by the Governor-General, Sir John Kerr, who installed Malcolm Fraser as prime minister on condition he call a general election. The election confirmed Fraser as head of a Liberal Government that was strongly in favour of mining and exporting uranium. South Australian Premier

Don Dunstan and Victorian ALP leader Clive Holding were the sponsors of a Conference motion to commit a future Labor Government to prohibit mining and export of uranium until the problems of weapons proliferation and waste management were solved.[8] The resolution, which was carried and therefore became ALP policy, also committed a future government to repudiate any export contracts approved by the Fraser Government.

The Fraser Government had announced its response to the Fox report in May 1977.[9] Basically it said that uranium could be exported, subject to a range of conditions. Broinowski notes that Malcolm Fraser made the extraordinary claim that the government had only decided to export uranium 'to strengthen Australia's voice in the moves against the proliferation of nuclear weapons'.[10] I have yet to find anyone who believed that statement. At the time, it was universally understood that the motive for exporting uranium was purely commercial. In any case, it was an odd argument that the best way to stop proliferation would be to export uranium, a bit like arguing that a reputable drug dealer who only sold high-quality cocaine would lift the standard of the entire illicit drug trade. The government set out a series of conditions that were meant to assure the community that Australian uranium would not be misused. Full IAEA safeguards were to be applied, uranium could only be sold to countries that had signed the Non-Proliferation Treaty, government-to-government safeguards agreements would have to precede commercial contracts, uranium could not be enriched beyond 20 per cent U^{235} or reprocessed without Australian approval and so on. Of course, to nobody's surprise, these stringent conditions were gradually watered down whenever they looked like impeding a commercial deal to sell uranium overseas. But first, there was the political issue of the ALP Opposition to deal with, and the opportunity to take advantage of the division within the union movement.

In 1977 the Liberal Party was in a Coalition Government with the Country Party, which has more recently renamed itself as the National Party. Its then leader and the Deputy Prime Minister, Doug Anthony, attacked the ALP position as coming from 'the wild men of the Left'.[11] He made the bizarre claim that not exporting uranium would be a 'clear breach of the Treaty on Non Proliferation of Nuclear Weapons'. In fact, the treaty imposed no obligation on its signatories to mine or sell uranium. To the contrary, it actually imposed stringent restrictions on those who did sell nuclear materials. Anthony also claimed that that there was 'no doubt' that the best way Australia could prevent weapons proliferation would be by 'developing and exporting its vast uranium resources as soon as possible'. The contorted logic was that other countries would be encouraged to build and use light water reactors if the market was flooded with uranium, but otherwise they might be tempted to build fast breeder reactors and create a plutonium economy. Again, this was a weird argument. I said at the time that it was like saying that we had to provide free seats on domestic flights between Sydney and Melbourne because people might fly around the world if we didn't.

An interesting proposal emerged in 1977 as a response to the concerns expressed by the Fox Report. I had been on a panel discussing the recommendations and suggested a possible model to allow Australia to export uranium with a clear conscience. Rather than selling to customers who processed the uranium for use in power reactors, trusting them to manage the radioactive waste that would result and prevent the use of fissile material as weapons, we could propose managing the entire process. We could turn the uranium into the fuel rods for nuclear reactors and lease them to users on condition they would be returned to Australia for final disposal. The representatives of mining companies on the panel were appalled and anxious to pour cold water on the idea. There would, of course, be complications, such as the need to negotiate with potential customers to produce the exact fuel rods they needed. I suspected that the underlying reason for their hostility was a belief that the fragile social license for exporting uranium would evaporate if we had to take responsibility for managing the waste.

Broinowski records that a group of Sydney lawyers actually made a formal proposal along those lines to the government.[12] They formed a company called World Wide Leasing Corporation. They said their scheme would ensure that safeguards applied to 'the length and breadth of the total nuclear fuel cycle' by ensuring that the uranium 'will at all times remain the lawful and legal property of an Australian-based company'. The proposal was summarily dismissed by the government. As Broinowski comments, the government stated its concern to prevent Australian uranium being used for weapons, but 'shied away from seriously considering a system that guaranteed that diversion could be avoided'.[13] He notes that the former Soviet Union employed this system to ensure its fissile material was not misused, but adds that 'the prospect of taking back nuclear waste … would present substantial political problems'.[14] I think that confirms my suspicion that the mining industry and governments believe that the community will support mining and export of uranium for the jobs and export revenue involved, but would not be willing to take responsibility for the radioactive waste produced.

As Cawte notes, there was bitter division within the union movement.[15] As head of the ACTU, Hawke negotiated an awkward compromise at their September 1977 Congress, calling on the Fraser

Government to hold a referendum on the question of whether to mine and export the uranium. Predictably, Fraser refused to hold a referendum, but decided to call an early election for December 1977, portraying his united government as more reliable than the bitterly divided ALP. He was returned to office, albeit with a reduced majority. Behind the scenes, the lofty safeguards were already being wound back.[16] The requirement for full IAEA safeguards was dropped on the practical grounds that these arrangements could not be applied to the yellowcake Australia planned to export. Then export contracts that had already been signed were deemed to be exempt from the condition that Australian consent would be needed for enrichment or reprocessing. Then there was pressure to relax the stipulation that uranium could only be sold to countries that had signed the NPT and were regarded as being responsible in their use of fissile material. France was regarded as a lucrative market, as the country had decided to close down its expensive oil-fired power stations by implementing a crash program of nuclear reactor development. Although Broinowski says 'Western intelligence' suspected the motives of the Shah of Iran, the Australian Government was still keen to sell his regime uranium.[17] The Australian Government was also centrally involved in the early stages of the messy situation that has since developed in the Korean peninsula.

Korea remains divided since the disastrous 1950s civil war, with a pro-US Government in the Republic of Korea, usually known as South Korea, and a communist regime in the Democratic People's Republic of Korea, usually known as North Korea. As Broinowski explains, the situation was still tense in the mid 1970s, with intelligence reports suggesting that South Korea was pursuing a clandestine nuclear research program.[18] They understandably feared they could be attacked by North Korea and thought the best way of ensuring the

United States kept enough troops in the country to repel any attack would be 'to threaten to acquire their own nuclear weapons'. This should have been at least an issue requiring serious concern and reflection, rather than a rush to sell uranium, but it wasn't. An agreement was signed in 1979 and South Korea rapidly became a major purchaser of Australian uranium, which it remains to this day. Then, in 1987, it was discovered that North Korea seemed to be developing the capacity to manufacture nuclear weapons, allegedly stockpiling plutonium from a small research reactor.

In the early 1990s, Democrat US President Bill Clinton negotiated an agreement to try to prevent the development of nuclear weapons in Korea. The deal was that North Korea would close down its research reactor, stop reprocessing its fuel rods to extract plutonium and adhere to the NPT. In return the United States would supply half a million tonnes of fuel oil a year as an interim measure, to supply North Korea's energy needs while two 1-gigawatt nuclear power reactors were being built. Unfortunately, the agreement unravelled when Republican George W. Bush was elected US President in 2000, prevailing over Democrat candidate Al Gore with a hotly contested win in the critical seat of Florida that had to be adjudicated by the US Supreme Court. Bush first slowed the delivery of the services Clinton had promised, then called North Korea a 'rogue state'. It responded by ordering the IAEA inspectors to leave and re-commissioning its research reactor. It seems clear now that North Korea has actually developed nuclear weapons. One of the few constructive actions of Donald Trump during his tempestuous term as US President from 2016 to 2020 was to reach out to the North Korean leadership with the goal of easing tensions, but at the time of writing it was difficult to be confident that anything significant had been achieved. While there remains a real risk that the tensions on the Korean peninsula could again boil over into armed conflict, it seems irresponsible to be tipping nuclear fuel onto the fire.

In September 1982, five months before the election that saw the ALP under Bob Hawke's leadership defeat the Fraser Government, the national parliament actually had a serious debate about uranium policy.[19] The ALP Opposition mounted a concerted attack on government policy. Lionel Bowen, their deputy leader, listed the failings of the safeguards regime. Uranium sold to Finland had been reprocessed in Russia, with no guarantee that plutonium was not being diverted to weapons production. Uranium was being sold to France, which had not signed the NPT and allegedly used the Australian metal for its power reactors to free up the uranium from other sources for its nuclear weapons program. The government was allowing sales to South Korea, which was suspected of having a weapons program, and planning sales to the unstable regime in the Philippines. He also attacked the irresponsibility of ignoring the problem of radioactive waste being produced from our uranium in overseas countries as 'out of sight, out of mind' and actually advocated the leasing proposal that had been floated in 1977, by which the government 'is bound to own and control the fuel ... gets it back, organises the question of reprocessing, plutonium storage and high-level waste disposal'. Broinowski rightly observes that it was easy to make that sort of proposal from the comparative safety of Opposition; when the ALP was elected only a few months later, the new government was just as reluctant as the Fraser administration had been to accept responsibility for the wastes that would inevitably be produced by nuclear reactors using Australian uranium.

The election of a Labor Government began a new period of political turmoil over uranium policy. Hawke and figures on the right of the party had begun at the 1982 Canberra national conference to water down the hard-line anti-uranium policy adopted in 1977, removing the commitment to repudiate contracts signed by the Fraser Government.[20] There was an emotional debate. Graham Richardson argued that closing the Ranger mine would lead to economic devastation, while Bob Collins reminded delegates that a group of working-class people had their livelihood dependent on the mine. On the other side, Stewart West reminded delegates that the problems that motivated the anti-uranium policy – proliferation, waste and local environmental issues – had not been solved. A proposed massive mine at Roxby Downs in South Australia became a critical issue. Western Mining Corporation (WMC) had identified a huge ore deposit that contained massive quantities of copper and significant amounts of gold, silver and uranium. WMC argued that the mine, a potentially major economic opportunity for the state, would only be viable if they were allowed to process and sell the uranium as well as the copper and gold. State Premier Don Dunstan had not been impressed by this argument. In one of his last acts before illness forced him to resign in 1979, he said, 'We simply cannot assure the people of SA that mining or treatment of uranium and the sale of uranium to a customer country is yet safe'.[21] The ALP lost the subsequent state election.

Under a new young state leader, John Bannon, it was seeking to return to power in the 1982 election. Hawke persuaded the conference delegates that the ALP would have no hope of winning the election if they were opposed to the Roxby Downs mine. The justification was that it was essentially a copper mine, with uranium, silver and gold as minor by-products. That was true, but the problem was that the minor associated production of uranium would make it one of the largest uranium mines in the world. The scale of the deposit is mind-boggling, estimated at about 3 billion tonnes of ore, of which about 1.2 per cent is copper and 0.04 per cent uranium, while on average each tonne of ore also contains about 6 grams of silver and about 0.5 grams of gold. It has been claimed that those figures make it the largest known deposit of uranium ore in the world and the third-largest known copper ore body. The ALP adopted what became known as 'the three mines policy', qualifying its general opposition to uranium mining by allowing the mines of Ranger, Nabarlek and Roxby Downs to go ahead.[22] Hawke laughed off journalists' criticism of the obvious double standards. Bannon won the state election.

The Roxby Downs mine began production in 1986 and a significant township has grown up around it and associated mineral-processing facilities. By 2005, it was producing 200,000 tonnes of copper a year and about 4000 tonnes of yellowcake, as well as significant quantities of silver and gold. BHP Billiton, which now operates the mine, estimates that about 70 per cent of the revenue comes from copper, about 20 per cent from uranium and the remainder from the smaller quantities of silver and gold. It is worth noting that the subsidiary uranium production is much greater than the average annual production of the Ranger mine, a project that precipitated a major environmental inquiry. Although the Roxby Downs mine is productive and commercially successful, its contribution to the broader economy has been much less impressive than the public were promised when the project was being considered.

In 1997, various groups in South Australia coordinated a public inquiry into the uranium industry.[23] The inquiry was co-chaired by Dr John Coulter, former leader of the Australian Democrats in the Senate, and Dr Tim Doyle, senior lecturer in environmental studies at the University of Adelaide. Its report set out a startling comparison between what had been promised in 1982 and what had been achieved fifteen years later. The mine was employing 946 people in 1997, compared with the 1982 claim that between 2268 and 2815 direct production jobs would be created. Cumulative royalties paid to the state in the decades 1988–2008 was estimated to add up to about $60 million, when over $2000 million had been promised (both figures in 1997 dollars). As the report drily concluded, 'the difference between the company's predicted benefits for the broader community and the actual benefits are quite profound'. That public inquiry actually concluded that the negative effects of the uranium industry on the environment and public health outweighed those modest economic benefits and recommended that the industry should be phased out 'as quickly as possible'. That recommendation had no political impact.

Successive governments, both ALP and Coalition, have supported the mine and even encouraged its operator to consider a plan for massive expansion. BHP Billiton conducted a feasibility study of a plan to transform the underground mine into an open pit operation, which would have created one of the largest holes on the Earth, about 5 kilometres in diameter and about a kilometre deep, increasing production of all the minerals by about a factor of five. State and Commonwealth governments appeared willing to fast track the necessary approvals for the huge project to go ahead, but BHP Billiton eventually decided not to proceed.

Hawke had declared when he was elected in 1983 that no new uranium export contracts would be allowed until the problems of waste management and weapons proliferation had been solved, but within a year his Cabinet had allowed new contracts for the sale of Ranger uranium. He also established in November 1983 an inquiry into Australia's role in the nuclear industry, fulfilling an election promise.

Professor Ralph Slatyer, chair of the national science and technology council ASTEC, conducted the inquiry and presented twenty-five recommendations to the government in May 1984.[24] The report was widely criticised. In a paper in the journal *Arena*, I pointed out that the report had conflated facts, such as the scientific knowledge about the possible leaching of radionuclides from borosilicate glass, with beliefs about the possibility of improving the technology.[25] It also made sweeping assertions about how the risk of weapons proliferation might be affected by the future sale of uranium on terms that had not been negotiated to countries that had not been identified. I argued that the same facts could easily have been used to justify totally different conclusions. Indeed, they subsequently were when the Total Environment Centre in Sydney convened an independent commission of inquiry into nuclear weapons and other consequences of Australian uranium mining.[26]

This inquiry led critics to conclude that the role of the Slatyer inquiry was to provide what looked like independent approval of the policies the government wanted to adopt. Indeed, when Hawke tabled the report in parliament, he described it as 'an independent and objective audit of policies and practices'. The report essentially said that Australia should continue to sell uranium, but to prevent weapons proliferation should only sell to those countries that had ratified the NPT and should promote peaceful nuclear technology. As Broinowski notes, while the report stressed that Australia should only sell uranium responsibly, it was not concerned about the successive steps that had gradually weakened the safeguards Fraser had announced in 1977.[27] I noted at the time that it was not just those critical of uranium export who believed the safeguards were ineffective. Similar views had been expressed by Sir Philip Baxter, Sir Ernest Titterton and the government's own

Uranium Advisory Council.[28] The assurances from ASTEC gave the impression of telling the government what it wanted to hear.

Ironically, the media cheer squad was using emotive language to contrast mining uranium as a matter of 'pragmatism and common sense' to avoid 'a dismal future as an energy-starved, backward country'. An editorial in *The Australian* in 1978 had said:

> The Australian people, as well as their leaders, must move away from making decisions in a lather of emotionalism whipped up by ill-informed or ill-intentioned propagandists. They must make decisions based on logical processes and accurate information, studied objectively and not through a mushroom-shaped cloud of hysteria and distortion.[29]

With no apparent sense of irony, that appeal to making decisions 'based on logical processes and accurate information' had suggested that Australia would face 'a dismal future as an energy-starved, backward country' if uranium was not mined, going on to equate opponents of the industry with 'those who feared the steam train'. By 1984, it was apparent that the slowing of the expected expansion of nuclear power after the Three Mile Island accident in 1979 had led to an oversupply of uranium, so the argument that an energy-starved world needed it had evaporated. ASTEC instead argued that Australia could 'contribute significantly to international energy security', the importance of which 'cannot be overemphasised'.[30] I thought it could be overemphasised and had been. There was no evidence that increasing the level of oversupply of uranium on a glutted market would do anything at all to improve energy security.

There were three other politically significant events in the 1980s in Australia. In the late 1970s, an outstanding scientist put forward a major innovation. Professor Ted Ringwood, chair of geology at the

Australian National University, proposed a very important step forward in radioactive waste management.[31] He had been concerned about the usual approach of nuclear authorities to embed the waste in a block of borosilicate glass for disposal in a stable geological layer. His concern was that glass is not stable over long periods. In fact, it is technically not even a solid but a meta-stable liquid and he did not believe it was safe to rely on it maintaining its integrity over thousands of years exposed to radiation and heat. He worried that hot groundwater could dissolve the radioactive minerals and disperse them into the wider environment. His insight was to recognise that some rocks do contain radioactive materials for geological time. That, after all, is what ore bodies such as those at Ranger and Roxby Downs represents, rocks containing uranium that have been stable for immense periods of geological time. So Ringwood proposed developing a synthetic rock, a concept for which he proposed the name 'synroc', to embed the dangerous radionuclides in the waste from nuclear reactors. He proposed a process that would cook rutile and titanium oxide with smaller amounts of three other minerals to fuse them together into a very stable synthetic rock. The effectiveness of his invention was obvious.

A group of my colleagues in the School of Science at Griffith University compared synroc samples with borosilicate glass and found the new material was orders of magnitude better at resisting leaching by hot water.[32] One observer commented that Ringwood found himself criticised from both sides. Some anti-nuclear activists were apparently dismayed that their key argument, the problem of managing waste, might have been demolished. On the other hand, those in the nuclear industry who had been assuring their politicians that it was quite safe to store radioactive waste in blocks of glass were not at all keen to admit there might be a better answer. Despite its obvious advantages, the nuclear industry globally has still not embraced synroc as a solution to the problem of waste management.

The Royal Commission chaired by Senator Jim McClelland was the second major event.[33] It recommended in 1986 that action should be taken to clean up the former British test sites at Emu Field and Maralinga 'so that they are fit for unrestricted habitation by the traditional Aboriginal owners'. A superficial clean-up had been done after the tests finished in 1967, but the Royal Commission found that several kilograms of plutonium had been left behind, some in shallow pits but much of it in fragments scattered widely around the site. The operation that was eventually conducted, only completed nearly fifteen years later, did not meet that object of unrestricted habitation. A government statement in 2002 acknowledged that there remains an area where people should not camp overnight, although they can travel through it for hunting and gathering. It said the nuclear regulator, Australian Radiation Protection and Nuclear Safety Agency, had found that the radiation levels were 'well below those that were anticipated at the start of the clean-up' but, as Broinowski observes, 'parts of the site will remain uninhabitable and dangerous for thousands of years from plutonium contamination'.[34] While the British Government contributed nearly half of the $108 million the operation was estimated to have cost, it is a distinctly unpleasant legacy of our colonial past that Australian governments allowed the testing of nuclear weapons to produce that situation.

The third event was the emergence of a single-issue political party, putting nuclear issues squarely into the electoral mix.[35] The Nuclear Disarmament Party (NDP) was founded in 1984 by Dr Michael Denborough, a Canberra doctor who was acutely disappointed by the ALP's move away from the anti-uranium policy that had been adopted in 1977. It had largely been overlooked, but former minister Don Chipp had given his disappointment about the Liberal Party's enthusiastic support for uranium mining and export as one of his reasons for resigning in 1977 to form a new political party, the Australian Democrats. He portrayed the new party as a middle voice between the Liberals and the ALP, much as the Centre Alliance founded by Nick Xenophon has done in more recent times. Chipp's party was successful in electing several Senators and often holding the balance of power in the upper house of parliament, giving it significant bargaining power.

Just as the Democrats represented a break away from the Liberal Party, the NDP had broken away from the ALP, with most of its members being disenchanted former Labor members or supporters. As Peter Christoff later wrote, the two issues that sparked the political move were the ALP decision to approve the Roxby Downs mine and the support for the military alliance ANZUS, especially for the presence in Australian ports of US nuclear warships. That question had been brought into focus when New Zealand Prime Minister David Lange announced that nuclear-armed warships would not be welcome in his country's waters. Since the US military had consistently refused to confirm or deny the presence of nuclear weapons on their warships, Lange's stance effectively banned the US navy from New Zealand. Hawke refused to align himself with Lange. The newly formed NDP had an explicit aim of removing Australia from supporting the US nuclear weapons system as well as stopping the mining and export of uranium.

Christoff observed that the new party was not at all well organised and had no coherent structure, 'a disparate assembly under one banner'. Its campaign for the 1984 election took off when it announced two star candidates for the Senate race, Midnight Oil founder and frontman

Peter Garrett in New South Wales and former ALP Senator Jean Melzer in Victoria. Christoff noted that the election campaign was marked by Garrett's 'articulate and persistent introduction of issues deflected, ignored or denied by the ALP'. He received 9.6 per cent of first preference votes in the election and Melzer 7.3 per cent. Overall, the NDP received over 600,000 votes, more than 4 per cent in every state except Tasmania, where it polled 3.9 per cent. Normally Peter Garrett's vote would have easily secured him a place in the Senate, but the ALP in New South Wales had decided to block him from being elected by directing its preferences away from him to the Coalition parties. Jo Vallentine, though, was elected as a NDP Senator for Western Australia.

Christoff notes that the overall vote for parties with 'a clear anti-nuclear platform' opposed to uranium mining, nuclear power and the nuclear military alliance was huge: about 1.1 million votes, 16.9 per cent of the vote in New South Wales, 15.8 per cent in South Australia and 14.3 per cent in Victoria. That should have sent a clear message to the governing parties that a significant minority of Australians were opposed to the support being given to the nuclear industry in general and uranium mining in particular. However, the rapid rise of the NDP and its lack of a coherent organisational structure brought it just as rapidly undone.[36] After a tumultuous 1985 conference in Melbourne, Peter Garrett, Jean Melzer and Senator Jo Vallentine resigned from the party, saying it had been taken over by a Trotskyist group, the Socialist Workers Party. Vallentine remained in the Senate as an independent 'senator for nuclear disarmament' and was re-elected in 1987. Without its star candidates, the NDP vote collapsed in the 1987 election, but through a quirk of the voting system their NSW candidate, Robert Wood, who had only received 1.5 per cent of the vote, was elected after distribution of preferences. In what became a long-running saga decades later, it was revealed that he had not been a citizen and was ineligible to be elected. A recount gave the Senate seat to the NDP's second candidate, Irina Dunn. After Wood subsequently obtained Australian citizenship, the NDP suggested that Dunn should resign and allow Wood to take her place in the Senate. She refused, left the party and joined Vallentine as an independent Senator for Nuclear Disarmament.

For the three years until the 1990 election, there were two senators with a single-issue focus on opposition to uranium export and Australian participation in the US nuclear military alliance. While there is little public concern about the mining and export of uranium in 2021, there is a significant movement worried about the military alliance with the United States in general and, in particular, the apparent willingness of Australian governments to follow meekly whatever ill-advised conflicts that country gets involved in.

In the late 1980s, an ambitious scheme was launched for an integrated nuclear industry in the Northern Territory.[37] Watters and Chandra, with support from the NT Government, published a discussion paper based on a presentation to the 1985 conference of the Australian Radiation Protection Society. It argued that the best way to minimise the risks of the nuclear fuel cycle would be to build on the Ranger uranium mine by locating near it facilities for enrichment and fuel fabrication, then adding reprocessing and waste management in the arid centre of the country. The enormous power demands of enrichment would have been provided by building a nuclear power station, while Ringwood's invention of synroc was proposed for storage and disposal of the radioactive waste. The integrated scheme would, they argued, allow Australia to effectively 'control the safety of much of the world nuclear cycle'.

At the time, the uranium industry was still confident there would be a dramatic expansion of nuclear power, from supplying about 10 per cent of world electricity to over 25 per cent by 1995, with associated trebling of the demand for processed uranium fuel elements. The proposal never attracted support beyond the Northern Territory, while even there some were understandably cautious about the scale of the upfront investment that would have been required. The expected expansion of nuclear power was stalled very soon after this proposal by the Chernobyl disaster in 1986, so it is probably a good thing the venture did not proceed.

That significant event in Europe will be recorded by history to have dramatically impacted the Australian uranium industry. The Chernobyl nuclear accident marked the end of political support for expanding nuclear power in Europe.

Chapter 7

CHERNOBYL, CLIMATE CHANGE AND FUKUSHIMA

The nuclear industry has a history of not being open about serious problems when they occur. There was a catastrophic fire in 1957 at Windscale in a UK nuclear reactor built solely to produce plutonium for the British bomb. I was told when I was a student that a physicist at University College London had detected an unusual level of background radiation in a laboratory one morning, but was puzzled to find the same anomalous reading outside in the street. He concluded that there must have been a nuclear incident somewhere nearby. It turned out that the incident was a fire in a reactor 500 kilometres away in the far north-west of England, in Cumbria, but the extra radiation had been detected in the streets of London. It was not at a level that was a health risk in the city, but it was a problem nearer the accident.

The fire and consequent release of radioactivity was hushed up by the Macmillan Government, realistically fearing it would erode public confidence in the nuclear industry and possibly harm their defence collaboration with the United States. A report into the accident was rapidly prepared but kept secret until 1988.[1] The Penney Report revealed that significant amounts of radiation had been released, particularly iodine-131, which is associated with thyroid cancer. The government was so concerned about this health risk that milk was confiscated from hundreds of kilometres around the site and tipped into the Irish Sea. The investigation found that the release of radiation would have been very much worse if there had not been a late addition to the construction of the reactor.

Sir John Cockcroft, leader of the design team, decided that a filter should be added to the building's chimney to trap radioactive dust if there were an accident. The device was dubbed 'Cockcroft's Folly' by other scientists, who thought the expense and construction delay were unjustified. It was estimated after the fire that the filters had probably trapped about 95 per cent of the radioactive material that would otherwise have contaminated the surrounding environment. Even so, subsequent analysis has concluded that at least 100 cancer deaths would have eventually resulted from the fire and the release of radiation.[2] The fire effectively destroyed that reactor. The government later renamed the site Sellafield, so the name Windscale no longer exists.

While the Windscale accident involved a reactor designed to produce weapon material, the civil nuclear power industry was comparatively safe until the 1979 Three Mile Island incident in the United States. A nuclear power station in Pennsylvania suffered a meltdown of its core. While there was no significant release of radiation into the surrounding environment, the accident seriously damaged the industry's credibility. During the 1977 debate in Australia about the nuclear industry, those who advocated building nuclear power stations quoted the Rasmussen Report, the result of a study by experts at the Massachusetts Institute of Technology for the US Nuclear Regulatory Commission.[3] The research used an approach known as fault tree analysis. It identified chains of events that could lead to a catastrophic accident, then calculated the probability of each event in the chain, and multiplied these to estimate the overall risk of the accident happening. This reached the comforting conclusion that the chance of a serious reactor accident was about one in a billion per unit per year. So even if there were to be 1000 nuclear reactors operating, we would only expect an accident about every million years.

That would certainly be an acceptable level of risk. The problem was, as I remember saying in the 1977 discussion, that the Rasmussen approach considered only identified failure risks whereas, almost by definition, accidents often occur because of risks that had not been identified. An obvious failing, as observed after the Three Mile Island meltdown by Canadian academics Joseph Morone and Edward Woodhouse, was that the fault tree analysis considered the risk of design errors or component failures, but human errors had contributed significantly to the accident.[4] Secondly, they said, regulators had concentrated on the risk of major problems, such as a failure of the cooling system, whereas the Three Mile Island accident was the result of a series of relatively minor malfunctions.

Similar conclusions often emerge from investigations of failures in other complex systems, such as aircraft crashes. There is rarely one single cause, accidents often resulting from a combination of causal factors, while human error is nearly always at least partly responsible. The Three Mile Island meltdown was a dramatic reminder that a series of unexpected minor malfunctions can be disastrous. It was also a reminder that human error can occur in complex and sophisticated technological systems. As David Collingridge observed, the distinctive problem of the nuclear industry 'is not the existence of misjudgement, wishful thinking, bad management, mistaken forecasts and plain bad luck, but the extraordinarily high cost which these errors have had'.[5]

It was also a demonstration that a warning issued in 1979, shortly before the Three Mile Island incident, was grounded in reality. The US Union of Concerned Scientists (UCS) had responded to what it called 'the official optimism about nuclear plant safety' by publishing a series of documents obtained by Freedom of Information requests to the country's Nuclear Regulatory Commission.[6] From a stack of reports about 30 centimetres thick, the UCS distilled a worrying summary of 'astonishing safety deficiencies' in US nuclear power stations.

It casts light on both the technical and institutional aspects of nuclear power reactor safety and shows how seriously the image of safety in the nuclear program is blemished by simple and widespread carelessness. For want of fuses, key nuclear safety equipment is rendered inoperative. Electrical relays fail because they are painted over or welded together or disconnected. Valves in safety equipment are destroyed because switches on their motors are incorrectly adjusted. A 3000-gallon radioactive waste tank is found in an operating plant to be connected with the plant's drinking water system. Emergency power sources fail routinely, bizarre equipment failure modes are a common occurrence and operator errors are an endemic feature of the US nuclear program.

The publication summarised the critical problem that sensitive pieces of safety equipment malfunction for a variety of reasons, including being so badly designed that a regulatory official had concluded they were 'guaranteed not to work'. UCS found it 'really distressing' that the regulatory agency which had catalogued the long list of defects had 'not taken adequate measures to prevent these recurring safety lapses'.[7] In that sense, Three Mile Island was an accident waiting to happen.

I was reminded of the detection of the Windscale fire in London when I read accounts of the April 1986 Chernobyl accident.[8] Monitoring devices in Sweden detected alarming levels of radiation. The scientists assumed there must have been an accident at one of their power reactors and quickly checked, but found nothing untoward had happened. We now know they were picking up the radiation reaching Sweden from an explosion and subsequent fire the previous day in a nuclear power station at Chernobyl, in the Ukraine – about 2500 kilometres away. The Soviet authorities had, like the UK ones in 1957, not made any public announcement as they scrambled to deal with the emergency. It was orders of magnitude worse than the British accident, releasing about 2000 times as much iodine-131, about 4000 times as much caesium-137 and about 500 times as much xenon-133, as well as large amounts of strontium-90 and significant amounts of plutonium. The World Nuclear Association (WNA) accepts that twenty-eight workers were killed, either by the explosion or by the acute radiation doses they received dealing with the accident, while there would have been at least 5000 extra cases of thyroid cancer among young people living in the surrounding area. The WNA correctly points out that the Chernobyl reactor that exploded during a test procedure, one of four reactors in a cluster at that site, was a Russian design with serious deficiencies that do not exist in typical Western designs, most obviously the lack of a containment vessel.[9] That being said, it is not clear that a containment vessel would have prevented the release of radiation from an explosion as catastrophic as the one that destroyed the Chernobyl reactor. Much has been written about the accident and there has even been a TV dramatisation, concentrating on the physical and emotional impacts on the humans who were centrally involved in the reactor design, the test procedure, the attempts to contain the fire and the bureaucratic responses. It is fair to say that the Soviet system does not come out of the incident well, but the British system did not come out well in the analysis of the Windscale accident either, and the US system was also shown by the UCS to be calmly overlooking obvious safety problems.

Although the Chernobyl accident was in a Soviet reactor with obvious design limitations, because the accident spread radioactive debris widely around western Europe, it had a devastating impact on the nuclear power industry more broadly. Morone and Woodhouse observed that Three Mile Island reminded people that something bad *might* happen, but Chernobyl showed that it *can* actually happen.[10] They quote a Harvard physicist, Dr Richard Wilson, as observing that the theoretical risk of a disaster had been realised. 'No one concerned with nuclear power, in the United States or elsewhere, can pretend that the Chernobyl accident makes no difference', he concluded.[11]

It certainly did make a difference. In several countries, planned expansion of nuclear power was put on hold or cancelled, while others moved to phase out nuclear power. Since the unification of Germany in 1989, no new nuclear power stations have been built, while those in the former East Germany have closed. Italy and Austria have voted not to have nuclear power and Switzerland to have no new reactors. However, the picture is certainly not uniform. France moved decisively in favour of nuclear electricity in the 1970s as a direct result of two historical forces. It had closed down its coal industry in the 1960s, replacing its coal-fired power stations by burning fuel oil. That looked a wise economic move at the time, with the world oil price less than US$2 a barrel. The OPEC oil crisis of the 1970s dramatically increased prices, which by 1979 reached about US$30. Suddenly oil-fired power was very expensive. France undertook a crash program of building nuclear power stations. Although there has been some retreat from the high point, about 70 per cent of French electricity still comes from nuclear power stations. The government aims to reduce this to 50 per cent by 2035 through expanding use of renewable energy systems, mainly wind and solar. Somewhat ironically, given the devastation of Chernobyl and the fact that a significant area surrounding the plant remains uninhabitable, the Ukraine has invested in new nuclear power stations since becoming independent and it now has one of the largest contributions from nuclear of any European country.

That being said, Chernobyl certainly marked a decisive turning point in the nuclear power industry, reinforcing a trend that Sverre Myrha detected after Three Mile Island.[12] The level of nuclear power began to decline gradually as old reactors were retired, while planned new ones were either deferred or cancelled. It looked like a dying industry in many countries. The UK Prime Minister Margaret Thatcher was deeply indebted to the nuclear power industry, because it had allowed her to successfully wage war on the union movement. She had been a minister in the Heath Government when it got into a major dispute with the union representing coal miners, in the early 1970s. When the government resisted the union demand for a pay increase, the miners went on strike. With the country's electricity system critically dependent on coal, a crisis ensued. To ration the limited power, the government put British industry on a three-day week. Finally, in desperation, they called a snap election, expecting the country to back their stand against the miners. To their shock, they were voted out. The incoming Labour Government negotiated a pay increase and the UK returned to normality. Mrs Thatcher certainly did not forget the incident.

Elected as Prime Minister in 1979, she was determined to bring the union movement to heel. She quietly approved expansion of nuclear power, and then lured the miners' union into a strike at a time when coal stocks were solid and there was plenty of back-up nuclear electricity.

After a long and bitter dispute, which almost bordered on civil war in the north of England, the miners were forced into a humiliating return to work.[13] The incident did not just critically weaken the coal miners' union, but also eroded the influence of the trade union movement generally. When the nuclear power industry was clearly in trouble after Chernobyl, Mrs Thatcher tried to help them out by legislating for a minimum share of delivered electricity to come from sources other than fossil fuels, ostensibly as a response to climate change. While the measure was designed to prop up nuclear power, instead it kick-started wind power in the United Kingdom.[14] From the viewpoint of the electricity industry, both construction time and capital cost looked better for wind turbines than large nuclear power stations, so that was where their money went. By the end of the twentieth century, nuclear power looked finished.

Then, as I wrote in a 2007 Quarterly Essay *Reaction Time*, something very strange happened. A small group in the United Kingdom hit on the idea of trying to rebadge nuclear power as the answer to climate change. This required a complete U-turn for most involved. For decades, they had fought a life-and-death battle with environmentalists, who had been consistently critical of the problem of waste management and, more recently after Chernobyl, attacking the industry because of the risk of local contamination from accidents. Desperate times call for desperate measures. The group of industry insiders suddenly aligned themselves with their traditional enemies, like Greenpeace and the World Wildlife Fund, embracing the science of climate change and briefing journalists with the new line: nuclear power is a relatively low-carbon source of large-scale secure electricity, whereas renewable sources like solar and wind are unreliable. The sun doesn't always shine, the wind doesn't always blow, but uranium can be relied on to continue

106

undergoing nuclear fission. Nuclear power is the clean alternative to coal for the baseload a modern society needs, was the argument.

At the international level, the argument got some traction. Consider the attempts to import this argument to Australia by John Howard in the dying days of his prime ministership, when his decade of studied inaction on climate change had become an obvious political liability. Where countries with a large land area and a small population, such as Australia or New Zealand, could easily imagine getting all their electricity from the emerging clean technologies with the right political will, it is less obviously a possibility for smaller and more densely populated countries like many of those of Europe. So some serious studies were undertaken to investigate the possibility of increasing the contribution from nuclear reactors. The main problem that became clear is one of economics.

As I previously noted, when I was living and working in the United Kingdom there was a vigorous debate about the relative cost of large-scale coal-fired and nuclear power stations. The issue was not clear, so the conclusion was critically dependent on the choice of assumptions about future fuel costs and the rate of discounting the future financial calculation. In the Open University's introductory materials science course, first offered in 1974, we included a case study of solar cells, then being used for exotic purposes like powering spacecraft. At the time, the panels were so expensive they could not be seriously considered for everyday use. We wrote at the time, before it was recognised that climate change was a serious problem, that the case for considering solar cells was the realisation that fossil fuels were a limited resource. The main material for manufacturing solar cells was silicon and sand is plentiful, so the use of solar cells looked a promising long-term alternative.

I became involved in the issue again in the 1980s as a member of the National Energy Research, Development and Demonstration Council. That body funded research at University of New South Wales that dramatically improved the efficiency of solar cells, producing more electricity from the sunlight hitting them. That trend has continued and the price of solar panels has kept coming down, so we now have a situation where literally millions of Australians have solar cells on their roofs. By 2007, when I wrote *Reaction Time*, the direction of investment was already clear. As energy markets liberalised around the Western world, investors turned their backs on nuclear energy. I noted that the number of reactors in the United States and western Europe had peaked about 1990 and declined since, with retirements and cancellations exceeding new orders. The figures for the decade up to 2003 for the average rates of increase in different forms of electricity supply were striking: wind power had increased by almost 30 per cent per year, solar by more than 20 per cent, gas 2 per cent, coal and oil 1 per cent, nuclear 0.6 per cent. The electricity industry was voting with its chequebook.

Despite the attempt to portray nuclear power as the technology of the future, the world was rejecting it in favour of alternatives that are cheaper, cleaner and more flexible. The figures also refute a claim sometimes made by those defending Australia's failure to take action to slow climate change. In 2007, it was argued that the 1997 Kyoto climate change targets were devised by cunning Europeans who could easily meet their emission reduction targets because they use nuclear power. Many European countries do have nuclear power stations, but I can only think of one – Finland – that has commissioned a new nuclear power station this century. Most actually have less now than they had in the Kyoto baseline year of 1990, so their nuclear electricity is not helping them at all to achieve their reduction targets. The small number of nuclear power stations still being built at the time of writing, in France, Finland and the United Kingdom, were all well behind schedule and well over budget. Three other problems remained for the dream of a nuclear renaissance.

Firstly, nuclear power is too slow a response to the urgent need to reduce the carbon dioxide emissions from the electricity industry. In fact, as discussed earlier in relation to the 1970s UK plan for thirty-six reactors, constructing a nuclear power station requires huge amounts of fossil fuel energy and large amounts of concrete, so building one actually increases emissions in the short term. In the long term, the energy investment is repaid, but it takes many years to build the power station and many more to erase the energy debt. By contrast, a solar panel or a wind turbine can be assembled in a few weeks and only takes one or two years to generate more energy than was needed to produce the device.

Secondly, climate change is actually reducing the usefulness of nuclear power stations. Increasing temperatures and changing rainfall patterns have reduced the availability of the cooling water they need. In recent summers, France has been forced to take nuclear power stations out of action because of this problem.[15]

Finally, the risks of the nuclear fuel cycle have not gone away. The Chernobyl accident was a tragic reminder that things can go wrong in power stations. There are now also the obvious risks of weapons proliferation or terrorism. Writing shortly after India had used peaceful Canadian technology to develop a bomb, the Fox Report warned that exporting uranium would inevitably increase the risk that other nations would develop nuclear weapons. Since then Pakistan and Israel have certainly developed nuclear weapons, it is widely assumed that North Korea has, and there are suspicions about Iran's nuclear program. The experience of Iraq, where the government was overpowered in a few weeks by the Coalition of the Willing coordinated by the United States, would have persuaded every tinpot dictator that you can be pushed around if you *don't* have weapons of mass destruction. The care with which world leaders tiptoe around Kim Jong-Un, recognising that North Korea probably does have nuclear weapons, will again have sent a clear message around the world.

When he was director of the International Atomic Energy Agency, Dr Mohamed ElBaradei told the 2005 conference reviewing the NPT of the vulnerabilities in the non-proliferation regime and his 'fears of a deadly nuclear detonation'.[16] He lamented the problem of trying to keep a beady eye on hundreds of nuclear installations around the world, on a budget comparable to that of a typical city police force. Despite his passionate pleas, for which his agency was awarded the Nobel Peace Prize, the UN conference ended in disarray, with the chairperson unable even to draft a final statement summarising the areas of disagreement. The fundamental problem is that most of the nations that have nuclear weapons are clearly in breach of the NPT, encouraging others to want to catch up.

China and the United States are both flagrantly developing and testing new weapons. As I was writing, the UK Government quietly announced the outcome of a major defence review. One of the policy decisions was to increase its stock of nuclear warheads from the current 195 to 260.[17] I contacted colleagues there and found none of them had seen any media reports of what would appear a major move, suggesting the government have not been eager to alert the voters to its decision. It is yet another example of the five nations that had nuclear weapons when the NPT was negotiated failing to honour their treaty obligation to reduce their stockpiles.

The deployment of nuclear weapons lends an extra dimension of instability to obvious global 'hot spots': the Middle East, the Indian subcontinent, the Taiwan Strait and the Korean peninsula. As discussed in the Introduction, a 2021 link-up of 25 Australian experts from a wide range of backgrounds identified nuclear weapons and accelerating climate change as the existential threats to human civilisation. Meanwhile, the problem of radioactive waste from reactors remains unsolved. Sweden and Finland have well advanced programs to develop secure long-term storage facilities, but the other countries with nuclear power stations are still just stockpiling the waste nearby, in most cases with no clear plan for disposal. I have argued that this is not just a technical challenge, but also a problem for our social institutions, because we need to devise systems that will remain secure for periods of time far longer than any human civilisation has ever endured. Not many of our leaders have got their heads around that dimension of the issue.

Since I wrote my Quarterly Essay in 2007, another catastrophic nuclear accident has happened.[18] The 2011 tsunami that overwhelmed the Fukushima-Daiichi power station really looks like the final nail in the coffin of the civilian nuclear industry. The cause was a massive earthquake, magnitude 9.0, about 100 kilometres off the east coast of Japan. The strongest earthquake ever to hit that country, which is no stranger to these events, it set off a tsunami of unbelievable scale. Waves were reported to have reached heights of 40 metres. While the Japanese coast has sea walls to protect it against this sort of disaster, those defences were ineffective against the onslaught of the ocean. Entire towns were swept away and it has been estimated that nearly 20,000 people died.

The Fukushima-Daiichi installation was a group of four nuclear power stations, located on the coast to use ocean water for cooling. A tsunami estimated to be 15 metres high flooded the site, interrupting both the main power supply that operated the reactors' cooling systems and the back-up generators. With no cooling, the reactors overheated and meltdowns effectively destroyed three of the reactors and severely damaged the fourth. Hydrogen explosions occurred and containment systems were breached, leading to 'the uncontrolled leak of radioactive materials beyond the vicinity of the plant'.[19] Comparison with the Chernobyl accident shows that the Fukushima event released about two and a half times as much xenon-133 and about half as much caesium-137, along with significant amounts of other radionuclides. The government ordered an evacuation of the surrounding area as a precaution, requiring about 100,000 people to move away. Because radiation is invisible, there was reported to be a high level of anxiety among those people, particularly the elderly.

While I was writing this book, a review to mark the tenth anniversary of the Fukushima disaster revealed the full extent of the problems facing the Japanese nuclear industry. Reuters Wire Service reported that 'an army of engineers, scientists and 5,000 workers are still mapping out a project many expect will not be completed in their lifetime'.[20] They have succeeded in removing the spent fuel rods from two of the reactors, but they now need to find a way to remove and safely store an estimated 880 tonnes of highly radioactive nuclear fuel as well as the concrete and metal that was contaminated when the fuel melted. Because it is not safe for workers to go into the area, the clean-up will need to be managed by robotic devices that do not yet exist but will have to be invented. There is no agreed plan for disposing of the heavily irradiated material when it is removed.[21]

'It's no good just moving highly radioactive waste from inside the nuclear reactor to somewhere else in the plant,' said Hiroshi Miyano, the head of the decommissioning committee of the Atomic Energy

Society of Japan. 'Where will the waste go?' he asked. That really is the 64-million-dollar question. A similar problem has arisen with contaminated water. About 1.2 million tonnes of the liquid is in temporary stores, holding the equivalent of about 500 Olympic swimming pools, but they are nearing the limit of the storages on the site. There has been speculation that it might be necessary to tip the radioactive water into the Pacific Ocean, but the local fishing community is understandably fighting that plan. While 'Japan's government says the job could run 40 years,' Reuters said, many experts think it could take up to twice as long.

The World Nuclear Association report of the disaster noted that the plants had been built about 10 metres above sea level to protect them against tsunamis, based on the government advice at the time of construction. However, it added, some eighteen years before the 2011 tsunami, new scientific advice suggested there was a risk of a major earthquake causing a tsunami up to 15.7 metres high. Their analysis continued:

> However, this had not yet led to any major action by either the plant operator, Tepco, or government regulators, notably the Nuclear & Industrial Safety Agency (NISA). Discussion was ongoing, but action minimal. The tsunami countermeasures could also have been reviewed in accordance with International Atomic Energy Agency (IAEA) guidelines which required taking into account high tsunami levels, but NISA continued to allow the Fukushima plant to operate without sufficient countermeasures such as moving the backup generators up the hill, sealing the lower part of the buildings, and having some back-up for seawater pumps, despite clear warnings.[22]

Takuji Hara of Kobe University has observed that a government seismologist had raised in 2009 his concern about the possible impact of earthquakes and tsunamis off the Pacific coastline of north-eastern Japan on the Fukushima Daiichi nuclear installation.[23] The warning was based on analysis of a magnitude 8.3 earthquake that is known to have struck the region more than 1140 years ago, triggering enormous tsunamis that flooded vast areas of Miyagi and Fukushima prefectures. That research concluded that the region should be alerted of the risk of a similar disaster striking again, but Hara notes it 'was not taken seriously by the authorities'.[24]

The subsequent analysis by Richard Hindmarsh concludes that the Japanese regulator, the Nuclear and Industrial Safety Agency (NISA), failed to enforce the sort of safety measures that should have been implemented, given that eighteen years had elapsed between the scientific advice and the disaster.[25] Others have written about the problem that has been called 'regulatory capture'. The agencies that should be regulating an industry often get into a cosy relationship in which they are reluctant to impose what are seen as unnecessarily costly safety measures. I remember when I was working in the United Kingdom, *The Guardian* newspaper exposed the scandal that the Avonmouth smelter had been releasing large amounts of lead, effectively poisoning nearby residents, but the regulator did nothing because the operator said that jobs could be lost if they were forced to upgrade the plant. Martin Frackler argues that 'collusive ties between regulators and industry led to weak oversight and a failure to ensure adequate safety levels'.[26] In the case of the Tokyo Electric Power Company (TEPCO), there was the obvious issue of not being prepared for the scale of the tsunami that occurred.

There was another important issue. After the Three Mile Island accident, the International Atomic Energy Agency recommended that nuclear power stations should install a venting mechanism so that any hydrogen produced by an accident would be harmlessly released into the air, rather than building up and risking an explosion. This is a potential problem when a reactor overheats because the zirconium cladding around fuel elements can react with steam to produce hydrogen. TEPCO never implemented that basic safety measure and NISA never required it. As a result, the accident was made considerably worse than it might have been by hydrogen explosions.

The whole debacle led the Swiss-based investment bank UBS to call the disaster 'the most serious ever for the credibility of nuclear power'.[27] While they accepted that 'Chernobyl affected one reactor in a totalitarian state with no safety culture', Fukushima raised a more fundamental issue, 'casting doubt on whether even an advanced economy can master nuclear safety'.

The accident caused an immediate shutdown of the entire nuclear power industry in Japan while safety was reassessed. It took years for the nuclear power reactors to be gradually brought back online. Atsuyuki Suzuki reports that 'stress checks were quickly performed' on many reactors in other countries.[28] The disaster also had a huge impact on public perceptions of the safety of nuclear reactors. A poll conducted in several countries in April 2011, the month after Fukushima, found a majority opposed to nuclear energy in all the countries surveyed except the United States, where there was a bare 52–48 majority supporting nuclear energy, and India where there was still strong support, 61–39.[29] The percentages opposing or strongly opposing were 51 in the United Kingdom, 58 in China, 59 in Japan, 62 in Russia, 66 in France, 79 in Germany and 81 in Italy. When those who indicated they opposed nuclear energy were asked if their opinion had been altered by the recent events in Japan, it was clear there had been a universal shift in opinion against nuclear energy, with the trend strongest in the countries nearest to the disaster, Japan and China.

The conclusion of the bank UBS was the most telling one.[30] Three Mile Island showed that there could be an accident that would effectively destroy a nuclear reactor, but with good safety measures the problem could be confined to the plant and not affect the community. Chernobyl showed that there could be catastrophic accidents that render whole regions uninhabitable for decades, but that could be discounted as the sort of thing that can go wrong in a totalitarian state with 'no safety culture'.[31] But Fukushima showed that there could be a disaster in a sophisticated modern state with advanced technological systems and a reputation for quality engineering. That really did raise the basic question of whether the risks of nuclear energy could be managed even by an advanced society, given the inevitable tendency for regulators to get into a cosy relationship with the industry they are charged with overseeing.

The Merkel Government in Germany reacted to the accident and subsequent drop in public support by moving to close down its nuclear power stations. Although some critics have argued that the decision will temporarily increase greenhouse gas emissions from the country's electricity system unless it imports more nuclear power from France, the German Government has maintained the commitment to phase out nuclear power in that country by the end of 2022.

Chapter 8

NUCLEAR POLITICS IN TWENTY-FIRST CENTURY AUSTRALIA

While Bob Hawke had been Prime Minister and Barry Jones Minister for Science, Australia had taken the issue of climate change seriously. Jones had sponsored a small agency, the Commission for the Future (CFF), to increase public awareness of issues arising from science and technology that could have significant public impact. After CSIRO had organised a major national conference that brought experts together to consider the possible impacts of climate change, Greenhouse '87, it negotiated an agreement with the CFF to hold jointly a major program to inform the community. I was Director of the CFF that year and heavily involved in the whole exercise, which included supplements in newspapers, a conference event in Melbourne linked to centres around the country and weekend meetings in those venues. At the insistence of Henry Rosenbloom, who runs Scribe books, I wrote a paperback book *Living in the Greenhouse*, summarising what we then knew about climate change and its likely impacts in Australia.

The Hawke Government organised an activity around the principle of Ecologically Sustainable Development, with nine working groups feeding into an overall strategy. The process culminated in the adoption in 1992 by the Council of Australian Governments of a

National Strategy for Ecologically Sustainable Development (NSESD).[1] According to the NSESD, Commonwealth, state and territory governments are committed to pursuing a pattern of development that does not reduce opportunities for future generations, that strives for equity within and between generations, that recognises the global dimension of our actions, that maintains the integrity of our natural systems and protects our unique biodiversity. I have observed that it would take a very generous assessor to detect any sign that recent governments, Commonwealth or state, even know that there is a National Strategy for Ecologically Sustainable Development, let alone accepting that it is the framework within which day-to-day decisions are made. The closest would be the government in power in the ACT at the time of writing, an alliance of the ALP and Greens, with a clear commitment to both social justice and environmental responsibility.

When the 1992 Rio Earth Summit was held and the Framework Convention on Climate Change was developed, Australia was led by an ALP Commonwealth Government. We contributed positively at the first Conference of the Parties, COP1, which developed the Berlin Mandate as the basis for reducing greenhouse gas emissions to slow climate change. Sadly, by the time of the second COP in Geneva, the recently elected Howard Government was taking a different approach. As a member of the Australian delegation, I cringed as our representatives used dubious economic modelling to claim that reducing emissions would do unacceptable financial damage to Australia. It was an outrageous argument. Having lived in the United Kingdom in the 1970s, I had seen European governments respond to the OPEC oil crisis by bringing in measures to reduce waste and improve the efficiency of using energy, everything from subsidising home insulation and improving public transport to setting mandatory efficiency standards for appliances and fleet vehicles. These measures had not been introduced in Australia, so there were easy cost-effective ways for Australia to reduce our emissions and play a constructive role on the global stage. Instead, we were regularly labelled the Fossil of the Day by those groups working towards an agreement to slow climate change.

The third COP, Kyoto in 1997, was even worse. When US Vice-President Al Gore arrived and told the meeting that he was there on the authority of President Clinton to tell the US delegation to negotiate flexibly, he was supporting the move for the conference to adopt legally binding emission reduction targets for all the affluent countries. I had been shocked, but not really surprised, when I attended a briefing before the Kyoto meeting. The leader of the Australian delegation told us what was being taken to Kyoto as our national position. It basically represented the ambit claims of the fossil fuel companies and energy-intensive industries like aluminium smelting as the Australian stance, saying we would only accept a Kyoto agreement if we were given a uniquely generous target, basically a licence to continue polluting.

When a late-night session of horse-trading produced a table of targets, most affluent countries had accepted commitments to reduce their greenhouse gas emissions by between 5 and 10 per cent. There were three exceptions. New Zealand and Iceland were seen as special cases because they were already getting all or nearly all their electricity from renewable sources, so they had less capacity to reduce than countries still burning coal, oil or gas. Australia was also given a target of actually increasing emissions. Worse was to come. When the final document was tabled, literally at about 3 a.m. after a long session of negotiating, the Australian delegation announced that they wanted to pursue their argument that land use change should be included in the agreement.

At a scientific level, it was a perfectly valid argument. The atmosphere does not notice whether the extra carbon dioxide is produced by burning coal or burning trees. But everybody in the great hall knew what was being proposed. In the Kyoto baseline year of 1990, huge amounts of land clearing were happening in New South Wales and Queensland. Including land use change in the agreement was increasing Australia's baseline by about 30 per cent, meaning our Kyoto target would be to increase our emissions by about 40 per cent when almost all the other affluent countries were committed to reducing. Since no other affluent country was clearing huge amounts of vegetation, the provision is known around the world as 'the Australia clause'.

At the media conference when the terms of the Kyoto Protocol were announced, incredulous European reporters bluntly asked the conference chair Raoul Estrada, 'How did you let Australia get away with this?' He replied in careful diplomatic language that his brief from the United Nations (UN) was to negotiate an emissions reduction agreement that included all the affluent nations so if it had happened, hypothetically, that one country which only accounted for 1.2 per cent of emissions had been holding out for an outrageously generous target, he might have felt obliged to accommodate them to make the agreement universal. Everybody knew what he was saying. One of the Australian observers gave a media conference with a paper bag over his head, to reflect his embarrassment at being identified as an Australian.

The approach was entirely consistent throughout the decade that John Howard was Prime Minister. In the late 1980s, when I wrote *Living in the Greenhouse*, there had still been some doubt about the science of climate change. It was agreed that human activity was changing the levels of greenhouse gases like carbon dioxide and methane in the atmosphere. CSIRO's Dr Graeme Pearman had set up a measuring station at Cape Grim, on the north-west corner of Tasmania, to test the claims by northern hemisphere scientists. His work showed the same pattern of increasing levels. From basic physics that was developed in the nineteenth century, it was expected that this would result in increasing global temperatures.

By 1987, the record was showing the trend in average temperatures that could be expected, but most cautious climate scientists were saying it was too early to say that the climate changes being observed were being caused by the increasing atmospheric levels of the greenhouse gases. The United Nations set up the Intergovernmental Panel on Climate Change (IPCC) to coordinate the science and prepare reports for the UN. By 1996 the debate inside the scientific community was essentially over. So when John Howard became Prime Minister, the science was clear: human activity was changing the amount of greenhouse gases in the atmosphere, this was changing the global climate, posing serious risks for human civilisation that demanded a concerted response. Under Howard, though, our government did nothing. When George W. Bush was elected US President and said he would not ratify the Kyoto Protocol, Howard said Australia would also refuse to ratify it, despite having been given a uniquely generous target. Much later, after he was voted out at the 2007 election, Howard said that he preferred to trust his instincts on a question like climate change rather than listen to the experts.[2] I thought that an appalling admission of irresponsibility.

Before the 2007 election, it was obvious to thoughtful people within the Liberal Party that Howard's approach was an electoral liability. Kevin Rudd had become leader of the ALP and described climate change as the greatest moral challenge of our time, promising to ratify the Kyoto Protocol and commit Australia to strong action on climate change. Howard's environment minister, the relatively progressive Malcolm Turnbull, tried unsuccessfully to persuade Howard to defuse climate change as an election issue by ratifying the Kyoto agreement.[3] He told his colleagues in blunt and unprintable language that he had been unable to persuade the prime minister to accept his argument.

Two years earlier, I had addressed the National Press Club, explaining to the media representatives why I did not believe nuclear power would be a sensible response to climate change. Some people asked me why I was bothering, with the Howard Government in deep denial about climate change and the nuclear option not being considered. I explained that I gave the speech because I believed it would eventually become impossible for the national government to continue ignoring the reality of climate change. I also feared that it would be consistent with John Howard's character and his approach to other problems to devise a noisy distraction, such as canvassing the option of nuclear power as a low-carbon energy source. My fears proved alarmingly accurate.

The nuclear power industry in Europe had attempted to reinvent itself as the low-carbon power source the world needed. That campaign had not reached Australia when I addressed the National Press Club, but a change happened in 2006. John Howard returned from a trip to Washington and announced, out of the blue, that Australia had to consider nuclear power.[4] He said in that press conference that we needed to 'calmly and sensibly examine what our options are'. He continued to say, 'when all the facts are in, we can then make judgements'. He argued that there had been 'very little debate' on the issue for twenty-five or thirty years 'because everybody said, oh well, you can't possibly even think about it'. He concluded, 'That's changed a lot'. I said at the time that it wasn't clear that things had changed a lot, but John Howard set about trying to ensure they did. He hastily put together a task force, described by comedian John Clarke as 'an independent group of people who want nuclear power by Tuesday'. The process was so rushed that the prime minister was not able give the names of the group the day he told the media it was being formed; they had to wait a few days for the full complement to be identified.[5] He did tell them it would be chaired by Dr Ziggy Switkowski, at the time also chair of the Australian Nuclear Science and Technology Organisation (ANSTO).

The group travelled extensively and interviewed leading lights in the nuclear power industry around the world before presenting in 2007 their report, *Uranium Mining, Processing and Nuclear Energy – Opportunities for Australia?* Just as the Fox Report had been thirty years earlier, it was hailed by the media as a green light for the nuclear industry: one headline saw it as showing Australia would have nuclear power within ten years.[6] One journalist rang me and asked if I had been persuaded by 'the rational argument' in the report to 'move beyond my emotional opposition to nuclear power'. I explained to the reporter that I felt my rational opposition to nuclear power in Australia had been strengthened by the analysis in the Switkowski report, since the group was clearly sympathetic to the idea of building nuclear power stations.

So what did the handpicked task force conclude about the economics of nuclear power in Australia? It said:

> Nuclear power is likely to be 20 to 50 per cent more costly to produce than power from a new coal-fired plant at current fossil fuel prices in Australia. This gap may close in the decades ahead, but nuclear power, and renewable energy sources, are only likely to become competitive in Australia in a system where the costs of greenhouse gas emissions are explicitly recognised. Even then,

private investment in the first-built reactors may require some form of government support or directive.[7]

In a later section, the report also said:

> In a world of global greenhouse gas constraints, emissions pricing using market-based measures would provide the appropriate framework for the market and investors to establish the optimal portfolio ... while carbon pricing could make nuclear power cost competitive on average, the first plants may need additional measures to kick-start the industry.

I explained to the reporter I found this endorsement of the use of nuclear power underwhelming. Basically the task force concluded that introducing carbon pricing, a move to which the government was ideologically opposed, would make coal-fired power more expensive and go some way towards closing the gap, but even with carbon pricing nuclear power would still not be attractive to commercial organisations.

Reading the full report made the economics look worse. The report cited US studies of the likely cost of electricity from new nuclear power stations in the range A\$75–105 per megawatt-hour (MWh). It later stated that building a nuclear reactor in Australia was 'likely to be 10–15 per cent more expensive because Australia has neither nuclear power construction experience nor regulatory infrastructure'. So that would have put the price of electricity somewhere in the range from A\$83–122 per MWh. For a comparison, the report quoted what were seen as the prices per MWh in Australia at the time: coal \$30–40, gas turbines \$35–55, wind or small hydro about \$55, solar thermal or biomass \$70–120, solar cells \$120. So nuclear power was, according to the Switkowski group, not marginally more expensive than coal or gas but at least double the price and probably more, as well as much more expensive than some forms of renewable energy. In fact, at the time of the report, even the most expensive forms of renewable energy came in at similar prices to nuclear energy.

The Howard Government and the commercial media enthusiastically acclaimed the report as showing that nuclear power was on the verge of becoming economical in Australia. It is difficult to see how anyone who had actually read the report and analysed the figures could come to this conclusion. They might have focused on one breathtaking leap of faith. Having cited the claimed US prices and conceded that nuclear power would be more expensive in Australia, the report suggested that the actual cost could be much lower, perhaps in the region of $40–65, 'if Australia becomes a late adopter of new generation reactors'. That remarkable claim assumed there would be spectacular improvements in the design of a class of reactors that has not yet been built anywhere in the world, roughly halving the present prices. That really would be a faith-based policy, given that no such improvements have been achieved anywhere in the industry's sixty-year history. The claim was also based on an obvious logical contradiction.

For Australia to be 'a late adopter of new generation reactors' and have the projected cost-effective power, we would have to wait until the reactors that had not been designed were actually built and commissioned before placing an order, which would take another ten to fifteen years to be built. In fact, fourteen years after the report was tabled, we are still waiting for the fabled new generation reactors to be designed and built, so being a 'late adopter' would have meant waiting for decades. But that projected economic viability was being used to suggest we should have started building nuclear reactors back in 2007 when the report saw the light of day. Had we done so, we would certainly not have been 'late adopters' and obtained the promised economic advantage. The cold, hard facts that could not be glossed over showed that there was no prospect of nuclear power being economically competitive in Australia, even if there were an emissions trading scheme or some other form of price on carbon dioxide emissions.

At about the same time, even more outrageous claims about the possible economics of nuclear power were made by a group at the University of Melbourne.[8] After conceding that recent construction of nuclear power stations had been marked by delays and huge cost blowouts, the report effectively asked gullible readers to ignore the history. It quoted a Westinghouse claim that its Advanced PWR reactor, the AP1000, would cost $1400 per kilowatt for the first reactor and fall to $1000 per kilowatt for subsequent years, with the new reactors 'ready for electricity production three years after first pouring concrete'. Actual recent experience in the United States involved construction costs more like $5000 per kilowatt and much longer completion times than the optimistic three years. That report swept to a conclusion that if the AP1000 could be built in three years at a capital cost of $1000 per kilowatt, it would 'provide cheaper electricity than any other fossil fuel based generating facility, including Australian coal power'. Of course, if the cost of future power stations was one-fifth of those recently built and the construction time was somewhere between a quarter and a half of those recently built, the economics would look a lot better. I said at the time that a parallel argument might be that you should ignore the fact that my score in my most recent round of golf was in the mid nineties, and indeed all my recent scores had been around that mark; I would only have to get my score down to the mid sixties to be a contender for next year's Australian Open, so I should be given government support for my entry.

The only realistic basis for assessing the cost of new power stations is the cost of others that have been recently built. At an Adelaide conference in 2007, the editor of the industry journal *Nucleonics Weekly* warned against unrealistic optimism. Mark Hibbs told the conference that Westinghouse had lost several hundred million dollars on a new reactor project in Finland, while the design for two proposed reactors in Taiwan was then only about 65 per cent complete, eleven years after signing contracts with two US firms.[9]

Despite the desperate attempts to portray the report as hailing 'a glowing future' for nuclear power in Australia, it did not capture the public imagination. It did little to counter the strong public feeling that would have been a serious political obstacle to building a reactor. In the run-up to the 2007 election, the Australia Institute conducted a mischievous exercise. They noted the criteria for selection of a site for a possible first nuclear power station. It would need to be on the coast to have a supply of cooling water. It would need to be easily connected to the electricity network. Ideally it should be near a major load centre, such as one of the capital cities. Based on those factors, the Institute released a short list of possible sites. The tsunami of panic among sitting members of parliament was wondrous to behold! They obviously did not believe that the community was eager to embrace nuclear power.

In the background, the debate about uranium exports was quietly continuing. Remembering Malcolm Fraser's 1977 claim that an 'energy-starved world' needed Australia's uranium, I had a sense of déjà vu in 2007 when a House of Representatives committee tabled a report entitled *Australia's uranium – Greenhouse friendly fuel for an energy hungry world*. It contained an updated version of Fraser's rhetoric, seeking to turn a grubby commercial deal into a moral virtue:

> As a matter of energy justice, Australia should not deny countries who wish to use nuclear power in a responsible manner the benefits from doing so. Neither should Australia refuse to export

its uranium to assist in addressing the global energy imbalance and the disparity in living standards associated with this global inequity.[10]

By this logic, it is claimed to be Australia's moral duty to help the poorest people in the world to have dirty and expensive energy that will leave them the enduring legacy of nuclear waste.

I believe the best way to help developing nations in our region have the sort of energy services that Australians take for granted would be to promote solar and biomass technologies that are both more appropriate to their needs and more likely to provide jobs and other economic benefits in Australia than expanding highly mechanised uranium mines. When I checked the figures, I found that uranium accounted for about 1 per cent of Australia's mineral exports, ranking with such metals as tin and tantalum. Perhaps more interesting, at the time I was writing this, sales of yellowcake in the last year for which figures were available had produced sales revenue of $734 million, while the revenue from sales of cheese was about $1000 million.[11] Given that the safeguards agreements arguably have more holes than Swiss cheese and radioactive waste is more unsavoury than an old gorgonzola, I would prefer Australia concentrated on exporting cheeses.

That is, in a sense, all ancient political history. John Howard did not just lose office in 2007; he earned his place in the political record by joining Stanley Bruce as only the second prime minister in Australian history to lose his own seat at a general election, when ABC journalist Maxine McKew won Bennelong from him. With the election of Kevin Rudd and his subsequent ratification of the Kyoto Protocol, political attention turned to ways to effectively reduce Australia's greenhouse gas emissions. Nuclear power did not enter the discussions about emissions trading, carbon pricing and renewable energy targets. However, there have been continuing issues around uranium exports and there was the contentious issue of whether the ageing Lucas Heights reactor should have been replaced.

As the original High Flux Australian Reactor (HIFAR) reactor at Lucas Heights neared the end of its useful life, after several upgrades and refurbishments over the nearly fifty years it had operated, the Australian Nuclear Science and Technology Organisation (ANSTO) lobbied the Howard Government for funds to build a replacement research reactor. It became a significant issue of science policy, because it was likely a new reactor would cost several hundred million dollars. ANSTO argued that a new reactor was essential for three broad reasons.[12] The primary case was that the Lucas Heights reactor was the source of radioactive substances used for medical imaging and cancer treatments. Unless it was replaced, the argument went, Australians would die needlessly because the substances needed to diagnose or treat their illnesses would not be available. Secondly, it was argued, without a working reactor Australia would lose its expertise in nuclear science and the capacity to apply that to the needs of industry. Finally, ANSTO said, Australia's capacity to participate in international negotiations about nuclear issues would be reduced without the 'hands-on expertise developed … from operation of a research reactor'. In particular, it was argued that Australia's regional seat on the board of governors of the International Atomic Energy Agency would be at risk if there was no longer a working nuclear reactor in the country, depriving Australia of its important role in global nuclear diplomacy.

Each of these claims was contentious, but not from the viewpoint of ANSTO. I still have the covering letter that came with ANSTO's glossy brochure, saying it was 'to ensure that commentators such as yourself have factual and verifiable information available'. It is certainly true, as that document produced for ANSTO in 1998 argued, that the need for medical isotopes is considerable.[13] The organisation was, at that time, producing about 350,000 patient-doses a year, or about 1000 a day. ANSTO argued that 'almost every Australian will benefit from a radiopharmaceutical during the course of their lives'. That looks statistically credible; if 350,000 doses were shared equally, it would take about seventy years for each Australian to get one. However, I suspect that some patients with serious health issues get more than one to make up for those like me who are yet to 'benefit from a radiopharmaceutical'.

The most important medical isotope in Australia, technetium-99, has a half-life of only six hours. The good point about that is that the radioactive substance decays rapidly after being introduced to a patient's body, so it doesn't remain a risk to them or those they live with. It is usually obtained from the decay of molybdenum-99, which has a half-life of sixty-six hours. That means it is an advantage to have the source of the radioactive substance in Australia, but not essential. A hospital in Perth would probably get its molybdenum within 24 hours of it being produced at Lucas Heights, but it could equally well get it within 24 hours of it being produced at Amersham near London or by the Nuclear Energy Corporation of South Africa. Broinowski argues that 'nuclear medicine departments in major Australian hospitals routinely buy their molybdenum generators and other radio-pharmaceuticals from overseas companies'.[14] He notes that there were not critical shortages when HIFAR was closed for routine maintenance or for refuelling, and he quotes a former ANSTO Chief Research Scientist, Professor Barry Allen, as saying 'the $300 million new reactor will have little impact on cancer prognosis'.

Allen's argument was that the new reactor would allow ANSTO to continue producing the isotopes that had been used for the previous ten or twenty years, most of which could easily be imported or produced using a much less expensive device called a cyclotron, but would not help with what he regarded as the next step in cancer treatment. Allen concluded, 'I think a lot more people will be saved if we can proceed with targeted alpha cancer therapy than being stuck with the reactor when we could in fact have imported these isotopes'. To press the point that it is feasible for Australian hospitals and medical imaging services to obtain the isotopes they now use from overseas, Sutherland Shire Council commissioned a study that showed 97.3 per cent of international flights to Sydney airport arrive within two hours of the scheduled time.[15] Jim Green's doctoral research found that the case for a new reactor in Australia for radioisotope production was weak, arguing for an alternative strategy that would have avoided both the capital cost of a new reactor and the inevitable consequent problem of managing the high-level waste it will produce.[16]

It certainly is true that replacing the research reactor would maintain one location for nuclear science research, but there are other centres of nuclear physics in Australia that could equally well have used the sort of resources that a new reactor demanded. Broinowski quotes Professor George Dracoulis, a nuclear physicist at the Australian National University, who says nuclear science in universities is starved of funds.[17] The ANSTO document emphasised that the replacement reactor would mainly provide a source of neutrons, allowing one particular field of research. It is true that neutron beams enable important research projects and ANSTO has been involved in an impressive range of industrial applications, although they are overwhelmingly oriented towards the mining industry. The case for the new reactor quoted an independent study that found ANSTO's 'recent research activities' had provided an estimated economic benefit to the mining industry 'of at least $100 million annually' and 'in the order of $25 million annually' to other industries.[18] I found the breakdown of those figures really interesting. I am not sure Australian taxpayers would be cheered to know that our nuclear science organisation is subsidising the mining industry to the tune of about $100 million a year, four times as much as it assists all other industries put together. All that being said, Dracoulis argued that the replacement of the research reactor was simply designed to give the impression that the government was concerned about Australia's capacity for nuclear science, but the limited opportunities it offers mean increasing numbers of scientists outside ANSTO will leave for the United Kingdom or the United States to continue their research.

The issue about Australia's regional seat on the International Atomic Energy Agency (IAEA) board is complicated. At the time of the debate about replacing the research reactor at Lucas Heights, Australia was one of the ten members of the board. The IAEA regulations provided that the board members should be from countries that are 'the most advanced in the technology of atomic energy including the production of source material'. That wording suggests it was probably Australia's role as the world's largest exporter of uranium that gave it a claim to be on the board, although it also would have greater capacity for nuclear science than any other country in the region, even if the reactor had not been replaced. As it happens, since this issue was resolved, the IAEA has clearly decided to expand its board. In September 2020 it announced the election of a whole group of new members, including Malaysia and New Zealand, to its board that now numbers 35 countries.[19] That certainly clarifies the point that countries don't need to have a research reactor to be on the board of the IAEA.

In the end, it was what a senior government bureaucrat called 'the whole health line' that was used to justify spending $320 million on a new reactor, the Open Pool Australian Light-water reactor (OPAL), designed and built by the Argentina company INVAP.[20] Work on the site began in 2002 and the OPAL reactor was commissioned in 2006. It is described by the World Nuclear Association as 'a modern, powerful and effective neutron source' that uses slightly enriched uranium. An interesting complication is that Commonwealth stated when approving the new reactor that a condition for the project to go ahead was that ANSTO have identified a satisfactory strategy for managing the radioactive waste that would be produced.

Uranium exports have continued, with no obvious evidence they were influenced by the rapid turnover of prime ministers since 2007. Policies addressing or ignoring climate change have been significant in several of the changes. Malcolm Turnbull was removed from the leadership of the Liberal Party twice by forces opposed to his comparatively progressive approach to reducing emissions. Kevin Rudd's indecision after the failure of the 2009 Copenhagen climate change conference and the rejection by the Senate of his Carbon Pollution Reduction Scheme was arguably the trigger for his being deposed by his parliamentary colleagues. When Julia Gillard led a minority Labor Government after the 2010 election gave no party a majority, her need to negotiate with Greens MPs and Independents culminated in a comprehensive package of measures to slow climate change, but Tony Abbott demonised the carbon price as 'a great big tax on everything' and undid most of his predecessor's good work in his tumultuous period in office. It is sufficient to note that none of the discussion about climate change since Howard have involved any serious discussion of nuclear power or the significance of uranium exports.

Arguably the most significant political issue affecting the nuclear industry in twenty-first century Australia was not an action of the Commonwealth but the decision by the South Australian Government to hold a Royal Commission with a wide brief to consider all aspects of the industry, from mining to radioactive waste management. That is a continuing problem.

Chapter 9

RADIOACTIVE WASTE – A CONTINUING PROBLEM

Since the HIFAR reactor was commissioned in 1957, Australia has faced the problem of radioactive waste that needs to be managed to prevent risks to human health and harm to the natural environment. As mentioned earlier, the spent fuel from nuclear reactors contains a toxic combination of nasty end products. Some are long-lived, so they need to be isolated from the biosphere for thousands of years. That is a real challenge.

There is also a more mundane issue of so-called low-level waste. Some of this is lightly contaminated items that were used in nuclear medicine, such as gloves and other protective clothing. There are also used devices containing radioactive substances, such as smoke detectors. Uranium mining leaves behind waste, piles of rock from which some of the uranium had been removed, as well as tailings dams, ponds containing wastewater that is lightly contaminated. Early uranium mines like Rum Jungle in the Northern Territory were just abandoned at the end of their life, leaving a toxic mess to leach slowly into the local water system. That was totally irresponsible. As the Senate Select Committee on Uranium Mining and Milling observed in 1997, 'tailings management is among the most serious challenges facing the uranium mining industry, industry regulators and their scientific advisors'.[1] A 1996 Senate inquiry into management of radioactive wastes recommended that a national storage facility for low-, intermediate- and high-level waste should be established. Some twenty-five years later, there is still no national facility for any category of radioactive waste.

The Fox Report set out the problem of the tailings that remain after uranium is extracted from the mined ore:

> This material contains all the radioactive decay products of the uranium, which were responsible for most of the radioactivity in the original ore. In the ore, these minerals were associated with a larger volume of non-radioactive rock. Milling converts all this material into a finely ground, more easily dispersed form.[2]

One of the decay products, thorium-230, has a half-life of about 76,000 years. It decays into the radio-toxic nuclides radium-226, radon-222 and radon decay products, which will be continuously produced in dwindling amounts until all the thorium-230 has decayed away. With the uranium extracted, the concentrations of radium and radon will eventually fall to insignificant levels. However, this will not occur for more than 100,000 years, during which time the hazard will persist.

Even though these decay products are not nearly as nasty as those in spent fuel from nuclear reactors, it is stating the obvious to say that this is a considerable technical challenge. As well as the radioactive minerals, the report noted that uranium ore also usually contains toxic heavy metals, such as lead and cadmium, usually as sulphides. Management of tailings needs also to ensure that these metals are not released into the environment.

So let me start with a summary of the broader problem of managing radioactive waste. As I mentioned in the introduction, I was a member from 2002 until 2016 of the Radiation Health and Safety Advisory Council, a statutory body that advises the regulator Australian Radiation Protection and Nuclear Safety Agency (ARPANSA). It met two or three times a year and at nearly every meeting some aspect of the problem of waste management was discussed. As a general definition, radioactive waste is radioactive material for which no use is envisaged. Management strategies vary according to the radionuclides involved, their levels of activity and the forms in which they occur, both physically and chemically. The international practice has been to classify waste as low level, intermediate level and high level, depending on the degree of radioactivity. While that system is broadly useful, there can be issues around the boundaries between the different categories, since there are no quantitative definitions.

The Nuclear Fuel Cycle Royal Commission established by the South Australian Government in 2015 devoted an entire chapter of its report to the management, storage and disposal of nuclear and radioactive waste. I was on the Royal Commission's expert advisory committee and am comfortable quoting some of the main conclusions in that chapter, with some accompanying commentary: 'Australia holds a manageable volume of domestically produced low and intermediate level radioactive wastes. The wastes result from science, medicine and industry ...'[3] When the report was written, about 4000 cubic metres of low level waste was stored around Australia, mostly consisting of about 2100 cubic metres of lightly contaminated soil stored in the Woomera Prohibited Area and about 1900 cubic metres of operational waste from ANSTO, stored at Lucas Heights. That site also holds about 450 cubic metres of intermediate level waste as well as about 400 kilograms of used fuel assemblies from the OPAL reactor. Some of the used fuel from the previous Lucas Heights reactor was reprocessed overseas and the resulting waste has also been returned to Lucas Heights.

The New South Wales Government has expressed concern about the continued storage of waste at that site, now that the suburbs have spread around it: 'The safe management, storage and disposal ... require both social consent for the activity and technical analyses to ensure that the waste is contained and isolated. Of the two, social consent warrants much greater attention than the technical issues during planning and development.'[4] The Royal Commission noted that there had been many examples overseas of failed processes that had concentrated on the technical issues. Without community support, the report said, projects usually failed, regardless of their technical merit and the actual risks that would have resulted. The NIMBY (not in my back yard) syndrome is alive and well; understandably, most people would prefer a potentially hazardous product to be in somebody else's back yard, rather than their own. Attempts to reassure the community by experts who have already decided a site are invariably met with resistance. The successful ventures involved the community before doing the technical work:

> Low level wastes ... require containment and isolation from the environment for up to a few hundred years. Intermediate level wastes need a greater degree of containment and isolation. The hazard posed by both kinds of waste reduces over time.[5]

> Low-level wastes from nuclear medicine and industrial applications typically contains radionuclides with relatively short half-lives, forty years or less. This waste usually does not need shielding to protect workers during transport and disposal, but best practice is to isolate it for a few hundred years until the radiation is similar to normal background levels. Intermediate level waste needs to be shielded to protect workers and ideally should be disposed of deep under the ground in stable geological formations.

> The federal government controls and manages most Australian low level and intermediate level waste... There appear to be advantages in term of managing long-term risks in a purpose-built, centralised facility.[6]

As noted earlier, the largest volumes of low-level waste (LLW) are contaminated soil from CSIRO research before 1970, contained in about 10,000 steel drums at Woomera, and ANSTO's similar quantity, also stored in steel drums in dedicated buildings at Lucas Heights. The Royal Commission argued that there have been 'many thousands of shipments of LLW in Australia without any accident resulting in harm to workers, the public or the environment', suggesting that the risks of transporting this waste to a centralised storage facility would be low. It argued that it would probably be cost-effective to have one site rather than maintaining the current system, which involves not just the two major locations of Woomera and Lucas Heights, but also a large number of sites managing much smaller amounts of waste.

To demonstrate the scale of the present activity, the report noted that radioactive waste is stored at seventy-eight different locations in South Australia alone. As discussed below, this issue of the benefits of a centralised location has been asserted consistently since the Commonwealth Government announced in 2001 its intention to create a national storage facility, but the public inquiry into one specific proposal revealed that no attempt had ever been made to quantify either the benefits of centralised storage or the risks of transport. If the transport risk is actually zero, as the SA Royal Commission argued, the benefit is clear:

> Many countries, including Finland, France, Hungary, South Africa, South Korea, Spain and the United Kingdom, have developed and operate purpose-built low-level waste repositories. These repositories handle volumes far greater than exist in Australia.[7]

It is fair to say that the technology for managing low-level waste is well established and facilities have been built in a variety of climates, many in conditions less favourable for long-term storage than can be found in Australia. The report went on to argue that 'there is no need for a perfect site' since it just needs to be adequate and properly engineered 'with sufficient barriers that, in combination, provide for long-term containment and isolation of radionuclides'. That is an important point, with overseas experience highlighting the importance of multiple barriers between the waste and the environment, to minimise the possibility of a breakdown releasing dangerous amounts of radiation:

> Key elements of the successful development of a low level and intermediate level waste facility are acceptance by society that it has an obligation to manage the waste it has created and compensation to communities that host facilities for the service they provide.[8]

The report reinforced the earlier observation that success was more likely to be achieved if the community accepted the project by adding the comment that acceptance was more likely when the community was compensated for the burden they were undertaking. It went on to argue that we are now benefiting from activities such as nuclear medicine that provide benefits, so we should take responsibility for managing the resulting waste so that 'an unfair burden is not placed on future generations'.

It went on to note that the Commonwealth Government was 'managing a process to identify a site' for a low level waste facility.[9] The language here is telling. I noticed during the debate about a proposal to store waste in South Australia that those promoting the project called it a waste repository, while those opposing the idea called it a waste dump! The report commended the recent development at El Cabril in Spain as 'an example of a modern, purpose-built surface facility that uses the multi-barrier approach'.

The history is a very depressing one. Back in 1997, the Bureau of Resource Sciences conducted a detailed analysis of possible sites.[10] It identified eight possible regions for a radioactive waste repository, then drew up a set of criteria for an ideal site: low rainfall, free from flooding, suitable geology, low population density, no known significant natural resources, reasonable transport access, groundwater not being used, not an area of environmental or cultural significance, not in an area where land ownership or control could compromise the ability of the government to manage the facility. Of course, one of those criteria poses fundamental difficulties in Australia, if an area has been declared of no cultural significance in ignorance of the feelings of local Indigenous groups. The Bureau concluded that the most suitable region in the whole of Australia was the Billa Kalina area in South Australia. That region includes both Woomera and Roxby Downs, so it could be said already to be holding significant quantities of low-level radioactive waste.

It was not really a surprise when the Commonwealth Government announced in 2002 that it would establish a low-level waste facility near Woomera in South Australia. While SA Premier Mike Rann was an enthusiastic supporter of mining and exporting uranium from the Roxby Downs mine and even urged its new owners, BHP Billiton, to go ahead with a plan for massive expansion of the mine, he sniffed the political wind and supported local groups opposing the proposal. The nuclear regulator, ARPANSA, held public hearings in Adelaide, for which I was one of the advisors. The Indigenous group that would have been affected came down to the city for the session and made it abundantly clear that they did not want radioactive waste on their land. More generally, the people of South Australia and their government were hostile to the site being chosen by the Commonwealth

Government and imposed on their state. While it was being asserted that a centralised facility would reduce the risk to the community from the existing situation where waste is spread around the country, no attempt at even a cursory risk analysis had been made.

After the public hearings, Rann took a proposal to the state parliament and an amendment was passed to the earlier *Nuclear Waste Storage Facility (Prohibition) Act* 2000, which had made it illegal to establish 'certain nuclear waste storage facilities' in South Australia. The state government then launched a legal action in the Federal Court to block the Commonwealth proposal. The Commonwealth Government glumly withdrew the proposal and the Prime Minister's department lamented 'the failure of the states and territories to co-operate with the Australian Government in finding a national solution for the safe and secure disposal of low-level radioactive waste'.[11] It went on to urge states and territories to take responsibility for their waste. Subsequently, ARPANSA sought a commitment from all those jurisdictions that they would 'adopt world's best practice in the management of radioactive waste'.

In 2004, the Commonwealth Government said that their intention had been to co-locate the low-level waste facility with the national store for intermediate-level waste. Since a group of scientific experts had ruled out the possibility of the intermediate-level waste facility being in South Australia, the commitment to co-location meant the low-level waste site could not be in that state either. The Commonwealth Government said that the examination of possible sites would commence immediately and 'be completed as a matter of priority'. Two years later, it announced that three sites in the Northern Territory were being considered and that the Indigenous traditional owners of Muckaty Station had expressed interest in return for 'appropriate compensation'.

Muckaty Station had been returned to the traditional owners in 2001. Acting on behalf of the traditional owners, the Northern Land Council nominated a particular area within Muckaty Station for a waste repository and the site was specifically named in the National Radioactive Waste Management bill presented to the Australian parliament in 2007. Then all hell broke loose. Many of the traditional owners said they had not been consulted, had not given their consent and would be unwilling to accept the proposal, regardless of how much monetary compensation was on offer. The government and the Northern Lands Council stuck to their guns, and so a senior traditional owner started legal action. The trial began four years later in the Federal Court. Two weeks into the trial, the Commonwealth Government agreed not to proceed.[12] So it was again back to square one.

A new process was begun and was still underway as this book was written. The Commonwealth Government effectively called for location offers for the waste repository. In February 2020, the government identified a farm at Napandee, near Kimba on the Eyre peninsula in South Australia, as the preferred site after a process of technical assessment and community consultation lasting about four years. The government says Kimba will receive a Community Development Package of 'up to $31 million', including a $20 million community fund 'to provide long-term support for the region'.[13] The same announcement said that other communities that were considered during the site selection process would receive $2 million in grant funding.

While the Commonwealth says the site has been agreed, there still remain a few problems for the proposal. The plebiscite of eligible Kimba voters found 55 per cent in favour. That is a majority, but well short of the 65 per cent benchmark the government had set itself to show broad community support. A second problem is the opposition of the traditional owners. The government refused to include the Barngarla request to be included in the local vote. So the Barngarla commissioned a postal ballot of the traditional owners and found none in favour. The Barngarla Determination Aboriginal Corporation, which organised the ballot, says it will 'take whatever steps are necessary' to stop the project going ahead.[14] While that raises the prospect of another legal battle, there also remains the legal problem of the state legislation prohibiting the storing of nuclear waste. It was not clear at the time of writing whether the Liberal State Government intends to repeal the 2000 law, which might be difficult as it no longer has a majority in parliament and the ALP Opposition seem to be supporting the traditional owners. Alternatively, the Commonwealth could assert that it has the power to override the SA legislation by saying that some of the radioactive waste is a consequence of the defence activities that are a clear Commonwealth responsibility. If it did that, it is unlikely their Liberal colleagues in the state government would challenge them in the High Court.

The proposal was referred by the Commonwealth to the Senate Economics Committee. A majority recommended accepting the draft legislation, but it was opposed by three members of the committee, Sarah Hanson-Young of the Greens, Rex Patrick of the Centre Alliance and ALP Senator Jenny Mc Allister.[15] Resources Minister Keith Pitt was reported as saying that Australia is 'one step closer to siting a facility to safely dispose of our low-level radioactive waste and temporarily store our intermediate-level waste, a process which has been ongoing for four decades'.[16] That comment revealed that co-location of a low-level waste repository with temporary storage of intermediate-level waste, which the Commonwealth said had been ruled out by scientific experts in 2003, has been quietly slipped back onto the agenda.

There remains the issue of the spent fuel and high-level waste from the new OPAL research reactor. As Broinoswski notes, when the Commonwealth Government approved the proposal for OPAL, it specified that work had to begin on proving the suitability of a site to store the waste.[17] However, that condition had been modified even before work began when the Executive Director of ANSTO, Dr Helen Garnett, told Broinowski that it was only necessary for ANSTO to have identified a *strategy* for managing the waste. He sought clarification from the regulator, ARPANSA, and was told there are three possible solutions to the problem.[18]

INVAP in Argentina, who built the reactor, could be asked to take the spent fuel rods back, reprocess them to extract the uranium and plutonium, and then return the high-level waste to Australia for disposal. A second possibility could be for ANSTO to negotiate with the French company that has reprocessed the fuel rods from the original HIFAR reactor to come to a similar arrangement with the OPAL fuel rods. Failing that, ANSTO would have to show the regulator that it had entered into an agreement with another company in a country that has a safeguards agreement with Australia. Wherever the fuel rods might be handled, there will remain the crunch issue; the return of the remainder after processing will require the establishment of a suitable facility for permanent disposal of high-level waste.

Given the four decades of difficulty getting a site for the comparatively benign low-level waste, it is impossible to be optimistic about the possibility of the government quickly achieving that. As Broinowski points out, even if the fuel rods are reprocessed to extract the remaining uranium and the plutonium, they will still contain a very nasty mix of fission products that will be dangerous for many thousands of years.[19] It is not clear that the community realises that Australia now has a legal responsibility to take back and permanently store an increasing amount of high-level radioactive waste. As discussed earlier, the only countries that appear to have a credible strategy for disposing of high-level waste, Finland and Sweden, have worked for decades to secure public trust in their processes. Other countries with accumulating amounts of high-level waste, notably the United States and the United Kingdom, but also including regional neighbours like Japan and South Korea, have no agreed solution to the problem.

An obvious fundamental issue emerged in the discussions about storing low-level radioactive waste. Every proposal put forward by the Commonwealth Government has been opposed by the relevant traditional owners. In every case, they have objected to waste being dumped on their traditional country. Since the whole of the Australian continent was occupied by various groups of Indigenous people for tens of thousands of years before Europeans set foot in the country, it is true that any proposed site for radioactive waste storage will be on the land of a group of traditional owners. That suggests that respecting the wishes of the Indigenous people is likely to rule out any proposed site, but waste is being generated and has to go somewhere. It is not at all clear how that fundamental problem can be resolved. Historically, governments have simply ridden roughshod over the wishes of the traditional owners, but the pressures for reconciliation and truth-telling have mercifully ended that tradition.

Proposals for Australia to go one step further, not just taking responsibility for the waste we are producing locally but also accepting some of the consequences of our selling uranium by offering to store permanently the resulting high-level waste, means some experts see the continuing problem of our regional neighbours as a potential economic opportunity for Australia.

Chapter 10

WASTE AND THE
SA ROYAL COMMISSION

Australia has historically been happy to sell uranium with minimal safeguards to satisfy the community that the uranium isn't being turned into weapons and no responsibility for the radioactive waste that is inevitably produced by the nuclear reactors fuelled by Australian uranium. There have been sporadic suggestions that Australia recognise the principle of product stewardship and take back the waste produced from uranium. A group of Sydney lawyers proposed a leasing program in 1977, Labor Opposition politician Lionel Bowen floated a similar idea in a 1982 parliamentary speech and the Northern Territory Government in 1985 supported investigation of a proposal for an integrated industry in their jurisdiction. Politicians in government have sniffed the wind and realised that the community is prepared to allow mining and export of uranium as long as they can avert their eyes from the consequences. Mining companies are similarly anxious to just sell the stuff. I can vividly remember the alarm shown by mining company executives in 1977 when I floated the idea that we might accept our responsibility for the consequences of its use.

In 1998, a proposal by a company called Pangea was leaked to the media.[1] It proposed setting up a site in remote inland Western Australia to take in the world's dirty nuclear washing. It argued that Australia has the attractive combination of both geological and political stability, with large areas that are sparsely populated and therefore ideal as a site to dump radioactive waste. It was an interesting pitch, partly a moral argument – since we were helping to create the problem and better placed than most to deal with it, we should assume responsibility at least for the waste resulting from our uranium – and partly commercial, suggesting that most countries with nuclear power had no plan to manage the waste, so there was an opportunity for Australia to set up a very profitable industry. Pangea argued that the proposal had a whole range of benefits. It would reduce the likelihood of uranium being turned into nuclear weapons and hence increase the chance of disarmament. It would take very nasty waste out of circulation and therefore reduce the risk of catastrophic accidents. It was even argued that it would both support the United Nations and strengthen our alliance with the United States, even though those might be seen as conflicting objectives. Broinowski writes that the proposal did not attract support from Australian politicians,[2] but I certainly remember retired Prime Minister Bob Hawke enthusiastically promoting the scheme, albeit without getting any obvious public support from those who were still in elected office. With little backing for the idea, Pangea apparently just closed its offices and left the country.

The idea of taking in nuclear waste from other countries came up again in South Australia in 2014. The story of the SA Nuclear Fuel Cycle Royal Commission is an object lesson in one of its central findings: that social consent is a critical issue when discussing the possible expansion of the nuclear industry. While the process followed by the Royal Commission was clearly best practice and its report was an exceptionally thorough document, its most contentious recommendation failed to achieve the level of social consent needed. The extensive and expensive process came to nothing, despite the enthusiastic support of the government of the day and the excellent work done by the commission.

When Rear Admiral the Honourable Kevin Scarce was Governor of South Australia in 2014, he made a couple of speeches about the state's involvement in the nuclear industry. He argued that South Australia was missing an economic opportunity by simply mining and exporting uranium. As many have done before and since, he advanced the general proposition that we should process some of our mineral resources and add value to them rather than exporting the raw materials. Even John Howard had expressed similar thoughts at one point, drawing the parallel with the way we exported wool and subsequently imported suits to suggest we were missing out on the opportunity to add value to our uranium by processing it here, perhaps adding an enrichment facility.[3] That had been seriously considered by the Fraser Government in which Howard had been a minister, but it never got further than discussions in smoke-filled rooms.

Kevin Scarce added an extra dimension to the argument, however, by suggesting the state of South Australia might consider going into the business of managing radioactive waste. This was entering a politically contentious area. When Mike Rann had been Premier of South Australia, he was very happy to preside over the mining and export of uranium from the Olympic Dam mine and appeared totally unworried by the consequent problem of huge volumes of tailings, even though this is a significant environmental management issue. The tailings constitute a large body of low-level radioactive waste. But when the Howard Government proposed establishing a national facility in South Australia for the long-term storage of low-level radioactive waste, Rann shamelessly joined the local opposition to the plan and led the campaign that blocked the scheme. So there was a history of the state ALP Government supporting the mining and export of uranium but being hostile to the idea that South Australia should store comparatively benign radioactive waste, items like gloves used in nuclear medicine and old smoke detectors.

Rann moved on and his successor, Jay Weatherill, saw thousands of jobs lost when the Australian Government withdrew its support for the local car industry. He was eager to promote new economic opportunities for the state before the forthcoming 2018 election. So in 2015 he invited Kevin Scarce to head a Royal Commission 'to inquire into and report on the potential to participate in four areas of activity … that comprise the nuclear fuel cycle'. Those four areas were extraction and milling of radioactive materials, further processing of those minerals, nuclear power and waste management. The political calculus required a very demanding timetable; commissioned in March 2015, Scarce was required to report by May 2016.

The Royal Commission was set up to be independent of the government and it was determined to follow an open and transparent process. They organised a small office in the centre of Adelaide and recruited a talented group of about twenty staff, managed by Greg Ward, an experienced public servant. I was invited to join their Expert Advisory Committee, along with the SA Chief Scientist, Dr Leeanna Read, and three nuclear experts: Dr Timothy Stone CBE, a visiting professor at University College London; John Carlson AM, the former Director of the Australian Safeguards and Non-Proliferation Office; and Professor Barry Brook of the University of Tasmania. My involvement along with Professor Brook gave the group an interesting balance, as we had been the two protagonists in the 2010 'flip book', giving the arguments for and against nuclear power in Australia.

Professor Brook was sceptical about the possibility of scaling up renewable sources like solar and wind to achieve change on the scale needed but optimistic about the promise of new types of nuclear reactors, whereas I was sceptical about the promised new reactors and optimistic about the developments I saw in solar and wind. Ten years later, I can claim to have been on the side of history, as CSIRO and the Australian Energy Market Operator released a report in 2018, finding that large-scale solar and wind are by far the least expensive forms of new power supply, even with enough storage to be what they called 'firm capacity'. It would take at least ten years to build one new nuclear power station and the cost of electricity from it would be much more expensive than the alternatives.

The Expert Advisory Committee commented on drafts of the Issues Papers produced by the commission, as well as critically reviewing the Tentative Findings, released in February 2016, and the May 2016 Report.[4] The commission also established a Socioeconomic Modelling Advisory Committee, to guide the development and interpretation of economic assessments, and a Radiation Medical Advisory Committee to assist in the interpretation of medical evidence about the health effects of radiation.

While the commission was working, it was on notice that any recommendation to increase involvement in nuclear activities would face some strong opposition. Just two months after the terms of reference were released, traditional owners from fourteen different communities met at Port Augusta to share their concerns. They released a statement calling on the Royal Commission not to recommend uranium mining, nuclear reactors or 'nuclear waste dumps' on their land. They joined conservation groups to form the No Dump Alliance in 2016.[5] The campaign was spearheaded by Yami Lester, a Yunkunytjatjara elder who was blinded as a ten-year-old by the Emu Field bomb tests. The group produced a postcard that read, 'South Australia, too good to waste'. Thousands of them were sent to the premier's office along with a petition. Altogether, it was reported that more than 35,000 people registered their opposition to further nuclear development.

Within a couple of months of being set up, the commission released four Issues Papers, one for each of the four areas being considered. A series of information sessions was held to raise awareness of the commission's work and invite submissions. More than 250 were received from individuals, community groups, professional organisations, corporations and industry bodies. A series of 45 public hearings then took oral evidence from 132 witnesses. The sessions were streamed live on the commission's website, where transcripts and videos were later available. Research was commissioned on the commercial viability and broader economic impacts of the possible activities being assessed. The commissioner and senior staff also visited a range of sites in Asia, Canada, the United States, the United Kingdom, Europe and the United Arab Emirates. In February 2016, a document entitled Tentative Findings was published and responses invited. About 170 submissions resulted. The commission considered these and produced its final report, published in May 2016 as demanded by its terms of reference.

It is a very impressive document, about 330 A4 pages of text and diagrams, tightly argued and extensively referenced. While the general criticism about objectivity set out in the Introduction applies and the commissioner himself was on the record before his appointment in support of expanding the nuclear industry in South Australia, the report fairly accepts many of the arguments of those opposing that view. Let me summarise the findings of the Report.[6]

It said that South Australia could 'safely increase its participation in nuclear activities'. While acknowledging that any participation has social, financial, environmental and health risks, it stated that these risks are manageable. It said that social consent would be 'fundamental to undertaking any new nuclear project', expanding that comment to point out that any such project would require new legislation, planning and the informed consent of the local community involved, as well as political bipartisanship and stable government policy, given the long timescales involved. With all those qualifications, the commission singled out the possibility that South Australia could offer to dispose of used nuclear fuel and intermediate level radioactive waste from other countries in the region. It estimated that such a waste management industry 'could generate more than $100 billion income in excess of expenditure' over a 120-year life. 'Given the significance of the potential revenue and the extended project timeframes,' it said, 'were such a project to proceed, it must be owned and controlled by the state government' with the wealth to be 'preserved and equitably shared for current and future generations'.

In addition to that headline finding, the commission made a whole series of findings in the specific areas it was asked to investigate:

- The existing requirements for approval of any new uranium mines involve duplication between the state and national government, so the process should be simplified 'to deliver a single assessment and approvals process'.
- The SA Government should invest in geophysical surveys in areas of high prospectivity with a view to identifying new uranium ore bodies.
- The full costs of future decommissioning and remediation of uranium mines should be secured in advance from mining companies.

- The concept of leasing nuclear fuel, rather than selling unprocessed uranium ores, should be explored and the existing legal prohibitions on processing should be lifted.
- The use of local facilities for producing medical isotopes should be expanded.
- While it would not be commercially viable to develop a nuclear power station in SA, the option of possibly using nuclear power as part of a future low-carbon electricity system should be kept under review.
- With that in mind, the SA Government should pursue removal of existing prohibitions of nuclear power, monitor new reactor designs and support development of a comprehensive national energy policy that would contribute a reliable and affordable low-carbon network.
- The SA Government should 'pursue the opportunity to establish used nuclear fuel and intermediate level waste storage and disposal facilities'.
- The SA Government should repeal the legislation prohibiting the storage of nuclear waste, to allow 'an orderly, detailed and thorough analysis of the opportunity to establish such facilities'.[7]

The proposal for South Australia to offer to store and dispose of radioactive waste needs further examination, but there are also a few other issues that deserve brief consideration. One is the proposal that the state government should invest in exploration of mineral resources, in the hope of identifying new sources of uranium ores. This presupposed that nuclear power could play an increasingly important role around the world as authorities seek to reduce the burning of fossil fuels, especially coal. That was a defensible view at the time the Report was written, though the report was suitably cautious. It actually said explicitly that predictions beyond 2030 are problematic. While there are some experts who think that emissions reductions consistent with the 2015 Paris Agreement targets will require much more use of nuclear power, there are other scenarios with little or no growth. 'Ambitious projections of long-term nuclear industry growth have a history of not being realised', the report warned.[8] That being said, the trends since the report was written have been very clear.

In 2019 and 2020 the world commissioned about 360 gigawatts of new renewables, mostly solar and wind with a smaller amount of hydro. The net new construction of nuclear power was almost zero, with the small amount of new capacity only slightly greater than the amount of older capacity that was retired.[9] With the world market for uranium oversupplied and little new nuclear capacity being commissioned, it does not appear a priority to invest in prospecting to identify new ore bodies. Even though the report was written five years after the Fukushima disaster, it did not fully appreciate the extent to which that dampened enthusiasm for nuclear power in countries with democratic governments. It is significant that the report warns explicitly about the need for concerted action to achieve the Paris target of well below two degrees increase in average global temperatures. Playing our part as a responsible global citizen will mean not just decarbonising the electricity system but also supporting pathways to reduce the climate impact of other sectors, especially transport. The report warned that it would be prudent to be prepared for the contingency of international pressure to phase out coal and gas more rapidly than Australian governments seem to intend. While the commission was not confident that nuclear power would be a useful addition to Australia's portfolio of low-carbon electricity supply systems, it suggested 'it would be prudent for it not to be precluded as an option'.[10] I can't personally see nuclear ever getting competitive with wind and solar, as their costs continue to come down, but I can't see any logical argument against keeping a watching brief.

Most importantly, the commission recommended that the SA government 'promote and collaborate on the development of a comprehensive national energy policy' aimed at producing 'a reliable, low-carbon electricity network at the lowest possible cost'.[11] The Gillard Labor Federal Government legislated, in consultation with the Greens and the cross bench, a package of measures that constituted a comprehensive approach to driving the electricity network towards a low-carbon future, but most of that package was irresponsibly torn up by the Abbott Liberal Government. Since then, there has been no attempt to have any sort of national energy policy, let alone a comprehensive one. Malcolm Turnbull's effort to implement a National Energy Guarantee was sabotaged by the far right of the Liberal Party, while Josh Frydenberg as his energy minister was attacking South Australia for investing in solar, wind and storage. Scott Morrison, before he deposed Turnbull to become prime minister, foolishly waved a lump of coal around in parliament to demonstrate his lack of grasp of climate change politics – it is an image that continues to haunt him. While South Australia clearly cannot shape a national energy policy on its own, and the Commonwealth Government appears totally incapable of providing the leadership needed, there could be an opportunity for the states to work together and circumvent the Canberra policy vacuum.

Without exception, the states and territories have adopted a goal of zero carbon by 2050. The ACT administration, ALP with Greens support, is taking serious action, while the Coalition Government in New South Wales has negotiated an ambitious low-carbon plan with the Greens and minor parties.

Another possibility raised in the commission's report is the idea of leasing nuclear fuel elements, rather than selling uranium.[12] While this seems an idea worth serious consideration, the uranium industry does not support it at all. I remember the representatives of uranium mining companies being appalled by my tentative suggestion of this in 1977, as well as the proposal by a group of Sydney lawyers and the 1982 speech by Lionel Bowen as an Opposition spokesman in the parliament. Nearly forty years later, when the SA Royal Commission floated the idea, the hostility of the current uranium mining companies was equally apparent. My interpretation of their reaction is that they believe there is a social licence to mine and export uranium because it provides a small number of local jobs and a limited amount of local revenue, with most people prepared to turn a blind eye to the long-term consequences of the industry: radioactive waste that needs to be safely stored and the problem of potential misuse of fissile material. If the inevitable consequence of selling uranium was to be the return of highly radioactive material for storage, I think the company executives are understandably worried that the social licence might evaporate.

A third major issue is the idea that we should repeal the existing prohibitions of using nuclear power and monitor development of new reactor designs, keeping alive the possibility that an improved or less expensive design might contribute to our need for a low-carbon electricity supply system.[13] I thought it was already clear five years ago, when the commission reported, which way the cost curves were going. Large-scale solar and wind are much cheaper supply options than any of the existing designs of nuclear power and the costs of solar and wind are still coming down, while it requires optimism bordering on delusion to see any realistic prospect of nuclear electricity becoming competitive. By 2019 the Australian Academy of Technology and Engineering, which has been a leading voice calling for use of nuclear power, said in its submission to the House of Representatives committee inquiry that developing a regulatory framework for nuclear electricity would be challenging and would 'consume valuable policy and regulatory design resources that might otherwise be dedicated to more pressing challenges in energy policy'.[14] It is difficult to argue that the prohibition on nuclear power should continue, since there does not seem any prospect of an electricity company wanting to spend their money on a nuclear power station or any jurisdiction being eager to licence one, but there is equally little enthusiasm among politicians for the possibility of a public brawl about repealing the existing legislation.

The Royal Commission also devoted considerable attention to the recurring idea that Australia should process its uranium before selling it.[15] There have been several attempts over the years to set up uranium enrichment facilities in Australia. The Australian Atomic Energy Commission had put enormous amounts of effort and resources into exploring the technologies that could be used to enrich uranium. There had been secret negotiations between the Commonwealth and the National Party's Bjelke-Petersen Government in Queensland in the late 1960s to spend a billion dollars setting up an operation near Rockhampton. Subsequently Rex Connor, the controversial Resources Minister in the Whitlam Labor Government, believed it would be a good idea. Connor enthused that enriching uranium would quadruple the economic benefits of exporting uranium. That proposal lapsed when the Whitlam Government was deposed, but behind the scenes the Fraser administration continued to discuss proposals with different states. A newspaper report in 1980 claimed 'Western Australia, Queensland and South Australia are all jockeying for the leadership in the enrichment stakes'.[16] The report went on to say that the Tonkin Government in South Australia had considered a joint project with the European Urenco consortium to spend A$500 million on an enrichment plant. The SA Government had apparently indicated that it was prepared to take up some equity in the project in the hope this would accelerate development of the state's uranium resources as well as adding value to the product. So it was entirely understandable that the recent Royal Commission would re-examine the idea.

Firstly, it concluded that there would be 'no technical impediment to providing conversion, enrichment or fuel fabrication services in Australia'.[17] In other words, it would be feasible for an Australian manufacturer to convert uranium to the gas uranium hexafluoride, enrich the gas to increase the ratio of uranium-235, and produce fuel rods for nuclear reactors. The obvious complication in the final step is that every type of nuclear reactor requires a specific type of fuel rod. While conversion and enrichment can be carried out at a large scale, fuel fabrication would demand specific retooling for each different customer. The Royal Commission noted that the technologies involved are sophisticated, but there was no technical reason why they could not be transferred to Australia. There are, however, legal impediments.

When the Howard Government legislated the *Environmental Protection and Biodiversity Conservation Act* 1999, political parties that held the balance of power in the Senate were opposed to further development of nuclear activities. The Act was amended to specifically prohibit the Commonwealth minister from approving enrichment or fuel fabrication. Ironically, the legislation, which has been widely criticised by scientists and environmental groups for its ineffectiveness and confirmed by the 2020 Graeme Samuel review to have done little to conserve biodiversity and protect the environment, was seen by the SA Royal Commission as preventing technologies that it saw as potentially providing cleaner energy.[18] The report noted another legislative impediment. The state law, the Radiation Protection and Control Act, also prohibits conversion and enrichment. In the case of the local law, there is an unusual provision that the prohibition 'may be removed by proclamation by the Governor'.[19] Since the commissioner was the former governor of the state, he would have been aware of this legislative quirk.

While the Royal Commission observed there was no technical impediment to further processing of uranium, as well as the legal barriers there is the commercial reality. The report found that 'the market for uranium conversion, enrichment and fuel fabrication services is over-supplied'.[20] It went on to add that the extent of the oversupply meant there was little prospect of new operators finding a demand for their services. The report noted that the demand for uranium processing was directly related to the number of nuclear power reactors operating. Shutdowns since the Fukushima accident had significantly reduced global demand. There was also an increasing trend towards nations having policies aimed at self-sufficiency, as exemplified by China's policy of handling its own needs for conversion, enrichment and fuel fabrication. The report also noted that some enriched uranium had been released from decommissioned weapons and military stockpiles. Adding together the legal barriers and the current commercial reality, there was no doubting the commission's conclusion that the recurring pipe dream of processing uranium in Australia will remain unfulfilled.

The most important finding of the Royal Commission was its conclusion that the SA Government should consider the possibility of establishing storage and disposal facilities for used nuclear fuel and intermediate level radioactive waste. Their argument was that there are currently large inventories of these wastes in 'safe but temporary storage'. At the time of the Fox Report, then prime minister Malcolm Fraser made the absurd claim that the waste problem had been solved. It hadn't, and more than forty years later, it still hasn't. As the commission report says:

> Used nuclear fuel, a solid ceramic in metal cladding, generates heat, is highly radioactive and hazardous. The level of hazard reduces over time with radiation levels decreasing rapidly during the first 30 to 50 years of storage, with the most radioactive elements decaying within the first 500 years. However, the less radioactive but longer-lived elements of used nuclear fuel require containment and isolation for at least 100,000 years.[21]

That timescale is truly mind-boggling. It demands not just technical precision but also, as mentioned earlier, appropriate social management, because the waste needs to be safely stored for a period that is orders of magnitude longer than any human civilisation has ever endured. In most countries with nuclear power stations, the used fuel is simply stored above ground near the power stations. As the commission report states, there is widespread agreement that the best long-term approach is deep geological disposal, such as Finland's and Sweden's well-advanced programs for deep underground disposal sites for their radioactive waste.[22] However, most other countries are still in a state of advanced political paralysis. In both the United Kingdom and the United States, proposals have been canvassed and opposed by local communities. The US Government's plan to store their waste at Yucca Mountain was abandoned when the State of Nevada finally withdrew its approval. The Royal Commission stated that the unsuccessful projects all began with technical plans that then failed to procure community acceptance, making the case that social consent would be essential for a waste management scheme to go ahead in Australia.

The report made the reasonable case that there are several other countries in the Asia–Pacific region, like Japan, South Korea and Taiwan, with the twin problems: significant amounts of used nuclear fuel awaiting disposal, and a difficulty finding geologically stable strata for possible disposal.[23] Based on preliminary discussions with officials in those countries, they were confident that a business case existed for South Australia to offer to dispose of wastes from those countries. Basically, with no local solution in sight, they would be prepared to pay to have their radioactive waste taken away for safe disposal. Excluding countries that are still trying to develop local solutions – United States, United Kingdom, France, Canada, Sweden and Finland – the report estimates that about 90,000 tonnes of used nuclear fuel is now available around the world for disposal, with a further amount about double that forecast to be produced between now and 2090 just through the continuing operation of reactors already built or firmly planned.[24]

The commission used a variety of measures to estimate what nuclear power utilities might be prepared to pay. Based on the estimated cost of advanced programs in Sweden, Finland and Switzerland, the funds set aside in some other countries and the costs of alternative approaches such as reprocessing, the commission concluded that it would be reasonable to calculate the finances of a repository based on a price of A$1.75 million per tonne of used fuel. It computed that a facility to store and then dispose of waste 'would be highly profitable' at that target price even if 'capturing only a relatively small share of the global inventory'.[25] The report estimated that it would take twenty to thirty years to construct the storage facility at an estimated cost of about $41 billion. The time and cost were assuming that a local operator would build on international experience in designing and constructing facilities for storage and disposal of radioactive waste.

Appendix J to the report gives more detail of the financial calcul ations.[26] Of about 90,000 tonnes of used fuel stockpiled around the world at the time, the holdings in Japan, South Korea and Taiwan added up to more than 40,000 tonnes. The cumulative forecasts, based on existing nuclear reactors, estimated that those three jurisdictions would account for about 112,000 tonnes by 2080. The Appendix also stated that those three hold 120,000 cubic metres of intermediate-level waste (ILW). As a basis for estimating the possible revenues from a waste facility, the baseline scenario went beyond the stockpiles in the Asia–Pacific region and calculated the returns if South Australia were to capture half of all the waste from nations not committed to local storage. It used figures of 138,000 tonnes of used fuel and 390,000 cubic metres of ILW, but also estimated a lower case of about half those amounts and an upper case of 50 per cent more. The price that customers would be prepared to pay is uncertain, as there is no existing market for those services. The report used what it described as 'a conservative price' of $1.75 million per tonne for storage and permanent disposal of used nuclear fuel. The estimates of capital and operating costs were 'based on costs observed and forecast for similar overseas facilities in advanced stages of development, converted to Australian dollars and scaled to the projected South Australian scenarios'. Both the likely costs and possible revenues were later debated, as the overseas figures vary considerably. The estimated costs of geological disposal range from $176,000 per stored unit in Finland to $1.3 million in Switzerland, so there was plenty of room for argument about the appropriate basis to compute the likely costs. Similarly, the fact that there is no existing market for the services means that there are different defensible estimates.

The main finding of the Royal Commission to the SA Government asked it to consider committing to spend more than $40 billion over twenty to thirty years and expect to earn more than $100 billion over the following century. Before the eye-watering sums spent by governments trying to keep economies afloat during the COVID-19 pandemic, those sums were big enough to cause any elected treasurer to go weak at the knees. While the proposal was sweetened by suggesting the future revenues should be invested in a State Wealth Fund, and one enthusiastic supporter claiming that this would bring untold wealth to the state, potentially allowing it to provide free electricity to everyone, others were sceptical of those promises. Understandably, the Weatherill Government did not rush to pull out the chequebook. Drawing on the key finding that any such initiative would require a clear indication of a social licence to proceed, the government decided to embark on an ambitious project of community consultation: a citizens' jury. This was entirely consistent with the Report, which said the government should 'facilitate a process of learning ... rather than conduct an exercise in advocacy and promotion' to improve 'the community's and government's understanding and awareness of the risks and opportunities'.[27] The Report thought it was feasible that 'the balance of informed public opinion could start to become clear after six to 18 months of engagement'. The commission made it clear that they believed the opportunities outweighed the risks and so they expected the 'process of learning' to lead informed citizens to the same conclusion.

The process used by the Weatherill Government could be seen as a model of public participation. Citizens were invited to volunteer for the role. From those who expressed interest, a citizens' jury of 300 was selected. They had access to the Royal Commission Report and all the supporting research. They spent three weekends together working through the ideas, exploring the risks and potential benefits. They were relatively free to decide which experts they wished to interrogate. In small groups they heard from those experts, questioned them and discussed the evidence presented, and then came back together to argue about the way forward. There were clear differences within the group, as would be expected. I gave evidence and answered questions. It was clear to me from the sorts of questions being asked and the tone of the discussion that some of the jury were in favour of the proposal and some were against it. As in any process of this sort, those with strong opinions seem to filter the evidence to reinforce their beliefs. Most of the jurors were undecided, genuinely seeking more information and trying to decide what would be right for the future of their state. While the Report had urged a process of learning, rather than advocacy or promotion, several of the jurors told me that they had been given clear indications the government wanted approval of the commission's recommendation to store radioactive waste. The jury was invited at least to support the idea of undertaking more detailed exploration of the proposal.

When it came to the crunch and a vote was taken, the citizens' jury rejected that possibility by a two-to-one majority. Essentially, a group of ordinary members of the community decided that the government should not spend more time and money exploring the proposal to store radioactive waste. Since the Weatherill Government had said clearly they would not proceed without public support, the jury had effectively scuttled the project. In their report, they made clear that the proposal foundered on the central issue of trust. The majority of the jury did not trust either the government or the private sector to construct and manage such a complex and important project. They also did not trust the financial calculations, which suggested that the waste repository would be an economic boon to the state.

The report by Daniel Wills in the Adelaide daily newspaper *The Advertiser* said:

> This 'bold' idea looks to have just gone up in a giant mushroom cloud. When Premier Jay Weatherill formed the Citizens' Jury to review the findings of the Royal Commission that recommended that SA set up a lucrative nuclear storage industry, he professed confidence that a well-informed cross-section of the state would make a wise judgement. Late Sunday, it handed down a stunning and overwhelming rejection of the proposal. Brutally, jurors cited a lack of trust even in what they had been asked to do and their concerns that consent was being manufactured. Others skewered the government's basic competency to get the thing done, doubting that it could pursue the industry safely and deliver the dump on-budget.[28]

I listened carefully to what members of the jury were saying about construction of the project. Their thinking was coloured by what was widely perceived as mismanagement by the state government of the contemporary construction of the new Royal Adelaide Hospital. The project ran over time, was well over its initial budget, and the final building was perceived to have serious shortcomings. If government could not manage construction of a hospital, they argued, how could they be confident they would get it right if entrusted with developing a facility to store highly radioactive materials? At least with a hospital they said, the government had previous experience, whereas nobody in Australia had ever built a repository for high-level radioactive waste and there was little overseas knowledge to draw on.

There was also no enthusiasm for the alternative of government managing private corporations to do the job. Some pointed out that the perceived failure of the Rudd Government's pink batts scheme, which led to young workers dying on the job, was due to the government entrusting the work to private corporations. The more recent experience of the Victorian government's problems with private security firms managing quarantine would reinforce the concern shown in 2016 by the SA citizens' jury. Some members also noted that the Fukushima nuclear disaster was much worse than it might have been because the Japanese Government had failed to regulate the private company operating the Daiichi reactors, allowing them to neglect obvious safety provisions that had been recommended by the International Atomic Energy Agency. So the majority of the jury did not trust the government to build such a complex project, but they also did not trust the government to manage private sector organisations with an obvious incentive to make a larger profit by cutting corners. They might have added that regulators sometimes don't have the knowledge or the political will to force the private sector organisations to comply with best practice.

The issue of the financial calculations also came down to trust. There is widespread cynicism about consulting companies, who are seen as providing their clients with the answer they want. Since the government was enthusiastic about the project, I heard some members of the jury say, of course the consultants did calculations to show it would be profitable. But they worried about the several unknown factors involved in the financial estimates. First, since nobody had yet built and operated such a facility, the capital and running costs could not be accurately known. What they did know was that most large-scale complex projects, hospitals or bridges or tunnels or stadiums or opera houses, turned out to cost much more than the public had been assured. So they were sceptical about the calculated costs of construction. Secondly, since there is no sort of market for the services of storing and disposing of radioactive waste, there was no way of being confident about how much the authorities in other countries might be prepared to pay. The Royal Commission had discussed possible pricing with some authorities and gave reasonable grounds for expecting the levels of payments they would be prepared to make, but there were no legal agreements about the possible revenue stream. There was also no way of guaranteeing in advance how much waste would be supplied to the facility. This was a classic chicken-and-egg dilemma; no responsible official in Japan or Taiwan or South Korea would commit to sending waste to South Australia until there was a commitment to build a properly engineered repository, but it would be difficult for responsible public officials in South Australia to commit at least $40 billion to build the facility without concrete assurances of the future revenues.

Finally, there was the question this raised of whether the project would be profitable and, if it were, if those profits would be sustainable. One member of the jury expressed concerns to me that the government could commit many billions to build the facility, then find that either the volumes of waste being sent there or the prices offered could mean that the project was not profitable. On the other hand, she said, if it really turned out to be as profitable as we are being told, why wouldn't another country jump in and offer a cut-rate service? Countries with much lower wage rates and less rigorous safety standards could presumably build and operate a repository for much less than it would cost in Australia, then undercut our market by offering to store the waste for a much lower price. So the majority of the jury basically decided that it would not be a responsible decision to go ahead with the proposal to set up shop as the world's radioactive waste centre, given their lack of trust in both the possible management of the project and the likely financial returns.

That was the majority view, but many of the citizens involved were strongly in favour of the project. It is also fair to note that the economists who provided evidence, including some who are strong supporters of environmental conservation, were solidly in support of the project, which they saw as a unique economic opportunity for South Australia, considering the state's small manufacturing sector had been devastated by the Commonwealth Government's decision to close down the car industry. The economists saw the nuclear waste management proposal as a new, technologically demanding industry that would provide a variety of ancillary jobs. Certainly the construction of storage and disposal facilities as well as the associated transport systems would have been a huge economic stimulus.

The discussions in the citizens' jury highlighted the problem of developing public understanding of complex technical issues. It reinforced conclusions from similar exercises relating to food irradiation and genetic modification of food crops. These have shown that educating a community group is very unlikely to produce consensus because we all have a tendency to see what we want to see. This is obvious at a mundane level if you talk to supporters of opposing football teams after a close match. In the case of educating a community group about a complex technical issue, it has been observed that those who start out uninformed but vaguely sympathetic usually became well-informed and sympathetic, while those who start out uninformed but vaguely hostile usually became well-informed and hostile. Most process the information they receive within the framework of their underlying values, unconsciously deciding whether to see the glass half-full or half-empty.

There is a more fundamental problem, identified by the late Lord May when he was UK Chief Scientist. In a presentation I witnessed at the 1999 UNESCO World Conference on Science in Budapest, he pointed out that most technical experts believe what has been called the 'deficit model'. As experts, they know that nuclear power or genetic modification of food crops is safe and effective. If the community were more scientifically literate, they believe, support would be universal. The problem, he said, is that the level of support for those technologies is highest in countries with the lowest levels of scientific literacy, the United States being the standout case with a significant fraction still thinking the sun goes around the Earth and an even larger fraction rejecting the theory of evolution. The level of support actually declines with increasing scientific literacy. It is lower in Australia than in North America, and lower still in western Europe. May's explanation was that only the uninformed believe the promises of unalloyed benefits from a technical innovation. Those who are better educated, he argued, know there is always a trade-off and realise that the benefits are always offset by extra risks or increased costs. This suggests that the process of education undertaken by the SA Government was always likely to lead to the group being better informed and therefore less likely to accept uncritically the assurances of the experts.

The Weatherill Government subsequently lost office. Stephen Marshall, when leader of the Opposition, showed little interest in the proposal and since being elected his government has shown no indication of wanting to revive the nuclear waste project. While the Royal Commission's work did not lead to the new industry that the commissioner strongly believed would have been a great opportunity for South Australia, the enduring legacy of the inquiry is a comprehensive report into all elements of the nuclear industry.

Chapter 11

NUCLEAR POLITICS
IN 2021

The nuclear issues of political significance in 2021 are electricity, submarines and weapons.

Given the lukewarm reaction to the 2006 UMPNER report and the panic among sitting politicians of that era when possible sites for a nuclear power station were suggested, I was disappointed to hear that the Morrison Government was setting up another inquiry. It asked the House of Representative committee on energy and environment to consider the obstacles to introducing nuclear power to Australia. Perhaps the prime minister was trying to deflect attention from his failure to respond to climate change, just as John Howard had in 2006. The committee was chaired by my local member in the electorate of Fairfax, Ted O'Brien. I had been puzzled when he erected a huge poster alongside a main road during the 2019 election campaign, promising cheaper electricity. Given electricity supply is a state responsibility, not a Commonwealth one, I found it difficult to see how his re-election could possibly lower the price of power. It was with a heavy heart that I prepared yet another submission.

The case I presented to the parliamentary committee explains why there does not appear to be any rational basis for expecting nuclear power to be used in Australia in the foreseeable future. It is too expensive and would be too slow to make a difference to our greenhouse gas emissions. Its water needs would be a fundamental issue on the driest of all inhabited continents. There would be almost intractable waste issues, given how difficult it has been to establish a site to dispose of even low-level nuclear waste. The perceptions of our neighbours could easily spark a nuclear arms race in our region. Finally, and most fundamentally, it is impossible to believe there would be community support for nuclear power.

In economic terms, the UMPNER report found that nuclear power would not be competitive unless there were to be both a price on carbon emissions and other financial incentives. In the absence of a carbon price, to which the current government is ideologically opposed, a very large public subsidy would be needed for electricity from a nuclear power station to be attractive to the operators of the Australian Energy Market. The Scarce Royal Commission had concluded that it would not be commercially viable to develop a nuclear power station in South Australia, even if there were to be unforeseen demand growth or strengthened inter-state connections to allow power to be exported to other states. Subsequent developments suggest strongly that the Royal Commission was optimistic in basing its calculations on the reported costs of Westinghouse's Advanced PWR reactor units being built in the United States. In 2017, Westinghouse filed for bankruptcy in New York, citing cost overruns, now estimated as over $15 billion on the four AP-1000 reactors it was building. Yes, the same AP-1000 reactors that the wild-eyed optimists at University of Melbourne were predicting could be built for $1000 a kilowatt and delivering power three years after the first concrete was poured. Bloomberg observed that 'Westinghouse Electric Co., once synonymous with America's industrial might, wagered its future on nuclear power – and lost'.[1]

Other recent overseas experience with nuclear power is equally bleak. The two French EPR reactors being built, one in France and one in Finland, are reported to be about $15 billion over budget, while it is estimated that the two UK reactors being built at Hinckley Point will eventually cost about $40 billion, with the project only continuing because the future operators have been guaranteed a return which is about double the present market price for electricity.[2]

By contrast, renewables continue to come down in price. The 2018 CSIRO–AEMO report found that the delivered cost of electricity from large-scale solar or wind was much less than the cost of power from any form of fossil fuel generation, even with enough storage added to be effectively firm capacity.[3] Those figures really are the killer blow for the prospects of nuclear power in Australia. Differences of 5 or 10 per cent might be possible to make up, but a factor of two or three is clearly right out of the question. Since the 2018 report, prices for large-scale solar and wind have continued to come down. A 2020 revision of the study showed that solar and wind power remain about $50 megawatts per hour, with the cost increasing to about $100 with enough pumped hydro storage to make them firm capacity.[4] The latest figures for gas were in the range $80–130, black coal $120–170 and brown coal $160–210, with no carbon price.

The report calculated that small modular reactors might have niche applications in Australia, but they are 'at least a decade away and would still deliver a levellised cost of energy at least twice that of wind and solar with storage'. In fact, the report gave the likely price of power from small modular reactors in the range $250–330! Even those figures are conjectural. The World Nuclear Industry Status Report 2020 said of small modular reactors, 'most designs are purely theoretical ones, and no real reactors have been constructed based on that design'.[5]

It concluded, 'Going by current trends, they are unlikely to ever be constructed beyond a few prototypes'.

To summarise the 2020 CSIRO/AEMO report:

> For generation costs, expressed as the levellised cost of energy ... wind and solar are clearly the cheapest form of bulk energy, at half the cost of fossil fuel alternatives, and one fifth the cost of nuclear. Even with storage – shorter duration batteries or longer duration pumped hydro – wind and solar match fossil fuels and are one third the cost of nuclear. In coming years, the costs of wind and particularly solar are expected to continue to fall. And storage too, so the cost advantage of renewables and storage is expected to increase.[6]

The really interesting conclusion of that latest report by CSIRO and the electricity market operator is the concession that large-scale storage is cost-effective. It confirms a landmark study by ANU scientists Andrew Blakers and colleagues, which showed that solar and wind with pumped hydro storage could provide all Australia's power needs at significantly lower cost than new large-scale power stations, either coal or nuclear.[7] They identified some 22,000 potential sites for pumped hydro installations around the national grid, with only the best fifty needed to provide enough storage to even out fluctuations in solar and wind energy, with a total water demand about 1 per cent of the volume currently used for irrigation schemes. By contrast, the demand for cooling water to operate a nuclear power station is huge, and an increasing problem with climate change.

Remember, nuclear power stations in France were taken off line in recent summers because of the shortage of cooling water. Given the state of Australia's inland rivers, the huge demand for cooling water to operate a nuclear power station means it would have to be sited on the coast to use seawater. There would almost certainly be environmental and cultural objections to any proposal for a nuclear power station on the coast of the eastern states. Previously, Professor Mark Diesendorf and his colleagues at University of New South Wales had modelled the hourly demand for electricity in the national system and came to the same conclusion, that all demand could be met by solar and wind with a modest investment in storage technologies.[8] The European Union has announced an intention to apply tariffs to products from countries that did not price carbon dioxide emissions or have serious policies to address climate change.[9] Of course, even a modest carbon price would make new fossil fuel projects even less economic, while having only a small impact on renewables or nuclear power.

Recent global data confirm that the economics of nuclear power are getting steadily worse, at the same time as the cost of large-scale renewable generating systems continues to come down. Germany's international broadcaster, Deutsche Welle, interviewed a leading world expert, Mycle Schneider, who edits the annual World Nuclear Industry Status Report.[10] He argued that nuclear power was not a useful response to climate change, 'not just because it is the most expensive form of electricity generation today but, above all, because it takes a long time to build reactors'. He gave the startling figures for changes in worldwide electricity prices between 2009 and 2020. The average price of coal-fired power increased marginally, from 11.2 cents per kilowatt-hour (kWh) to 11.2. Nuclear power, on average, increased from 12.3 to 16.3 cents per kWh, a 33 per cent jump. By contrast, wind power came down from an average price of 13.5 to 4.0 cents kWh, while solar did even better, improving from 35.9 to 3.7 cents per kWh.

Where solar power had been nearly three times the price of nuclear in 2009, by 2020 it was less than a quarter of the price of nuclear. Even gas looked good economically by comparison with nuclear, coming down over the same period from an average price of 8.3 to 5.9 cents per kWh. Schneider added that those figures are averages and some of the latest technology is even better, citing a case of solar power in Portugal being delivered at a cost of 1.1 cents per kWh, so cheap that even with storage it is less than just the operating cost of nuclear power stations. Schneider says that many of the reactors still running are no longer profitable, but are still in use because of accounting considerations. While a power station is operating, even if it is running at a loss, it appears on the accounts of the energy company as an asset. When it is closed down, the company doesn't just lose an asset from the positive column of its balance sheet, it incurs a liability – the cost of decommissioning the reactor, typically of the order of $1.5 billion. In Japan, he said, 'it often took years to officially close nuclear power plants because companies could not afford to remove [them] from their assets. Some of those operators would have gone bankrupt overnight.' Schneider believes that the French nuclear energy company EDF faces a serious financial crisis, having not put aside the funds that will be needed to close ageing reactors.

Even if nuclear power were a cost-effective response to climate change, it would not be a timely response, as pointed out in the 2007 UMPNER report. There has been no revision of its estimate that it would take at least ten years to build one nuclear power station in Australia, with fifteen years probably being a more realistic timescale. That means it would be well into the 2030s before a potential nuclear power station would first deliver a kilowatt of electricity if a decision to build one were taken this week. The late 2030s is well after the date by which significant reductions have to be achieved under the 2015 Paris Agreement. UMPNER also estimated that even a concerted move to nuclear power, building 25 reactors by 2050, would only reduce the growth in our greenhouse gas emissions by 18 per cent. Since it is clear Australia needs to aim for zero emissions in the future, with all states and territories now having accepted the goal of zero emissions by 2050 at worst, a reduction in the rate of growth is clearly not an acceptable outcome. So-called Generation Four reactors and small modular reactors have been promoted by some, but these are still on the drawing board and there are no credible forecasts showing they would be cost-effective power sources.

All this discussion centres on the need for electricity supply, but at some point the question of demand should enter the debate. The National Framework for Energy Efficiency, a report presented to the Howard Government nearly twenty years ago in 2003, observed that by far the most cost-effective way to reduce carbon dioxide emissions is to improve the efficiency of turning energy into the goods and services we want.[11] I remind people of the famous saying by US energy analyst Amory Lovins, who said the people don't want energy, they want hot showers and cold beer! Much of the technology we use to turn energy into services is still woefully inefficient. Appliance efficiency labelling was introduced against the stalling of the Commonwealth Government when ALP administrations in New South Wales and Victoria legislated for this basic level of consumer information, but many of the appliances on sale in Australia could not legally be sold in the European Union. Some of the worst devices dumped on the Australian market do not meet even the minimum standards in some states of the United States! Our buildings are among the worst in the developed world for energy efficiency, with the recent spread of unsuitable designs and thoughtless orientation leading to dramatic increases in the need for air conditioning to maintain comfort.

When I returned to Australia in 1980, about 5 per cent of dwellings in south-east Queensland were air-conditioned. The traditional Queenslander, on stilts with verandas and good cross-ventilation, is well suited to the sub-tropical climate. The spread of brick veneer boxes has necessitated a huge investment in cooling, with about two-thirds of dwellings now air conditioned. Improving the efficiency of using electricity makes more sense than increasing supply, whatever generating technology is used. But the supply will need to be enhanced if there is a serious commitment to a zero emissions future.

After electricity, the big-ticket item is transport. Several European countries have now set dates beyond which new petroleum-fuelled vehicles will not be allowed. The competing possible technologies, electric vehicles or hydrogen fuel-cell propulsion, will both require dramatic increases in electricity supply. There needs to be forward planning for this transition in Australia. The recent trend to increasing use of large four-wheel-drive vehicles to cope with the difficult terrain of suburban streets has cancelled out all the recent efficiency gains in engines, tyres and drive trains, so the average fuel efficiency of the vehicle fleet is about the same as it was fifty years ago. At least for city driving, the future will inevitably mean smaller, more efficient vehicles. It is technically absurd to use cars weighing 1.5 to 2 tonnes to move a payload typically less than 100 kilograms.

The move towards the renewable power sources of solar and wind is gathering pace. While Liberal Commonwealth ministers criticised the South Australian ALP Government for recklessly encouraging solar and wind power, as well as large-scale storage, the Liberal administration that replaced the Weatherill Government has quietly continued the same policies. The Coalition Government in New South Wales in 2020 negotiated with Greens and minor parties to implement a progressive policy of cleaner electricity; their road map shows 90 per cent of power coming from solar and wind with pumped hydro storage by 2030. In 2021, the Queensland Government said it would build five regional-scale storage batteries to make better use of solar energy.

By 2019 the Australian Academy of Technology and Engineering had accepted the improbability of nuclear power being used in Australia and effectively urged the Federal parliament to devote its policy capacity to more urgent priorities. In 2021 the Climate Council published a report showing that 7 gigawatts of renewables had been installed in Australia just in the one year of 2020, when much normal economic activity had been precluded by the COVID-19 pandemic.[12] As a result of the extra capacity, more than 30 per cent of all electricity delivered through the Australia national grid in the summer of 2020–21 had been from renewables, with less than 5 per cent coming from gas. Those figures exposed the absurdity of the argument the Commonwealth Government had been running, that increasing use of gas would be essential because of the unreliability of solar and wind. The Australian Government's absence of a coherent energy policy in 2021 was not just ignoring the science and out of touch with community opinion, it was also out of touch with its own branches in at least three states. To further back this up, in 2021 the former Australian Chief Scientist, Dr Alan Finkel, published a Quarterly Essay, *Getting to Zero*, on the challenge of getting to no emissions by 2050.[13] He argues that the goal is achievable, but it will require coherent and coordinated policies.

Another significant issue in 2021 is the question of whether future submarines should be nuclear powered. A one-day seminar was held in Canberra in 2019 to discuss the topic and a new Discussion Paper on the subject was released in 2021. The author was Lindsay Hughes, Senior Research Analyst in the Indo-Pacific research program at Future Directions International, an independent research organisation based in Perth. His paper argues that defence needs would be better met if the next generation of submarines were to be nuclear-powered rather than using conventional diesel-electric engines. It is worth quoting his summary:

> Australia badly needs a modern submarine fleet to replace its ageing Collins-class submarines in order to ensure the security of its sea lines of communication in an increasingly volatile Indo-Pacific region. Its future submarines will be required to operate further from home, including in the South China Sea.[14]

Accordingly, Prime Minister Tony Abbott announced early in August 2015 that his government would bring forward defence projects worth billions of dollars, including the Attack submarine (known at the time as the Future Submarine) project. In his estimation, doing so would safeguard thousands of Australian shipbuilding jobs and keep a large part of the estimated $40 billion that the proposed twelve new submarines would cost in the country. His announcement reinvigorated debate on whether the Future Submarines ought to be bought off the shelf from a foreign manufacturer, designed specifically for Australia's require-ments and built in conjunction with a foreign manufacturer or completely designed and built in Australia. The debate almost automatically assumed that the Future Submarines would be conventionally powered – that is, by diesel-electric engines.

A Defence White Paper of 2009 had dismissed the nuclear option as expensive to purchase and operate, requiring new skills and training, gaps in self-reliance and safety concerns. Left unstated was the fact that anything nuclear is politically dangerous to the government that proposes their acquisition and use. Nevertheless, in light of the delays and cost overruns that reportedly plague the Attack submarines, it may be time to revisit the nuclear-powered option and ask if Australia should also develop a nuclear power sector. To be clear, this is not to argue that Australia should create a nuclear-power sector in order to support a fleet of nuclear-powered submarines – there is no immediate connection between the two – but, rather, to show that Australia needs a nuclear sector and, if one were established, support a nuclear-powered submarine fleet that could better serve its defence requirements.

There is an interesting logical jump in the last sentence of the summary of Hughes' paper. It sets out the advantages that nuclear-powered submarines have over those using diesel-electric engines. They are quieter and therefore more difficult for enemies to detect, as well as being able to remain under water for longer. It quotes a retired US navy officer as saying that a single nuclear-powered submarine is effectively equivalent to between 2.2 and 6 conventional ones. The paper also points out problems with the decision by the Abbott Government to purchase the French-designed Shortfin-Barracuda submarines, which are 'yet-to-be-fully designed – let alone tested – conventionally-powered variant of a nuclear-powered submarine'. Hughes says the cost of the project has already blown out from the originally-quoted $50 billion to an estimated $89 billion, and says the delays being encountered have 'reportedly led Prime Minister Scott Morrison to contemplate terminating the entire project' – the prime minister being presumably involved because the minister for defence was on extended sick leave. Hughes goes on to lament the fact that the nuclear option is still not being considered.

I can see the logic of arguing that nuclear-powered submarines are operationally superior to those using diesel-electric motors. The US navy made that switch decades ago. However, I confess I am uneasy about the whole idea of putting nuclear reactors in places where other people will try to destroy them. I can't help worrying about the radiological consequences if a nuclear submarine were to be destroyed by enemy action. I am also not convinced at all that spending close to $100 billion on submarines will make Australia safer. Since the 1960s, Australia's military has been almost constantly engaged in fighting ill-advised wars started by the United States. It is not at all clear that Australia's national interest was served by trying to prop up a puppet regime in the south of Vietnam and there was even less justification for us helping to fight American wars in Afghanistan and Iraq. A 2021 public inquiry was being conducted by the Independent and Peaceful Australia Network (IPAN) into the costs and consequences of Australian involvement in US-led wars.[15] IPAN is a network of community, faith and peace groups as well as trade unions and concerned individuals. The justification for Australian submarines seems to be, in Hughes' words, 'to ensure the security of its sea lines of communication in an increasingly volatile Indo-Pacific region'. So that seems to be saying that we need submarines to ensure we can continue to export coal and iron ore. I wonder if the Australian community would agree that it is worth spending close to $100 billion so that mining companies can securely export their products to China and Japan.

Even if there is agreement we need to have submarines patrolling the South China Sea, I am still puzzled by the logical jump in the argument for a nuclear power sector to support that activity. The paper makes the obvious point that coal-fired power needs to be phased out to slow climate change, but without considering the other options concludes 'a sufficiently large nuclear power sector ... could drastically reduce Australia's greenhouse gas emissions'. The author goes on to argue that having a nuclear power industry 'would have the additional benefit of providing skills that could be transferred into the military domain – including nuclear-powered submarines'. That seems to be the hinge of the argument, leading to the conclusion that 'the creation of a nuclear-power sector ought to be revisited' because it 'could potentially provide much of the foundational skills required to maintain and operate a nuclear-power submarine fleet'. That really is the military tail wagging the electricity industry dog. But the influence of the defence sector is always strong, so the issue will continue bubbling along beneath the surface.

Of course, the issue of nuclear weapons is implicit in the comment that a nuclear power industry would provide skills 'that could be transferred into the military domain'. There is an increasing level of discussion about the rising power of China in the Asia–Pacific region and the declining influence of the United States. Some of the conversation implicitly revives the argument advanced by Sir Philip Baxter after the end of the American war in Vietnam, when he argued that Australia could no longer rely on the US nuclear arsenal to protect us against potential enemies, so we should acquire our own nuclear weapons.

The third big political issue in 2021 was that continuing spectre of nuclear weapons and the Australian Government's attitude towards international moves for nuclear disarmament.

Chapter 12

AUSTRALIA
AND NUCLEAR WEAPONS

There have from time to time been suggestions that Australia should develop its own nuclear weapons. Australia was involved in supplying uranium to the United Kingdom and providing test sites to help the development of the British bomb. One of the two main concerns raised by the Fox Commission about uranium exports was the risk that the fissile material could be used to develop what are now often referred to as weapons of mass destruction. That is an accurate label. While the sort of bombs that have conventionally been used in warfare could destroy entire buildings or production facilities, the nuclear weapons that have been developed since the 1950s could destroy entire cities. Even worse, there is credible research that a conflict that led to a series of nuclear explosions could put so much material into the atmosphere that the entire world could be plunged into a 'nuclear winter'.

Throughout my adult life, we have lived with that Sword of Damocles hanging over our heads. At times of international tension, there has been real alarm. I vividly recall the crisis in October 1962, when the US Government discovered that Russian missiles were being installed in the island state of Cuba. Although US weapons pointed at Russia had been installed in Europe, the Kennedy administration decided that the very idea of Russian missiles just across the Gulf of Mexico was an unacceptable threat. The world held its breath as Russian ships steamed towards Cuba and nuclear war seemed imminent. I heard students agonise about whether there was any point studying for the end-of-year exams. Some young couples reportedly brought forward marriage plans, worried that civilisation might be wiped out. The crisis was averted when the Russian leadership backed off and turned their ships around.

When scientists who had worked on the World War II bombs decided in 1947 to turn their internal communication newsletter into a formal journal, the *Bulletin of the Atomic Scientists*, they created a cartoon of a stylised clock to draw attention to the peril humanity was facing. It became known as the Doomsday Clock.[1] Its annual revision is a reminder to politicians and the public that we are all in mortal danger. The magazine's founders said the clock symbolised the urgency of the nuclear dangers that they and the broader scientific community were trying to convey to the public and political leaders around the world. The clock was set at seven minutes to midnight. Two years later, with the news that a nuclear weapon had been tested by the Union of Soviet Socialist Republics (USSR), the communist state centred on modern Russia, the clock was moved to 11.57. In 1953, after the United States first successfully tested the hydrogen bomb and the USSR followed a few months later, the clock was advanced to 11.58 with a warning there was a real chance that 'from Moscow to Chicago, atomic explosions will strike midnight for Western civilisation'.

Then there was a period of modest progress. It gradually became apparent that the new weapons were so powerful that only a deranged leader would consider using them against a similarly armed enemy, given the inevitability of catastrophic retaliation. In 1963, after they had been continuously testing more and more deadly weapons, the United States and the USSR signed the Partial Test Ban Treaty, which prohibited atmospheric testing. The clock was moved back to 11.48. It was a false dawn. The two super-powers simply shifted their testing of new weapons to underground facilities, while other countries such as the United Kingdom, France and China developed their own nuclear arsenals.

The clock gradually moved closer and closer to midnight until the mid 1980s, when it stood at 11.57. Then Mikhail Gorbachev assumed the leadership of the USSR and began a series of negotiations to ease tensions and reduce the risk of nuclear war. The fall of the Berlin Wall in 1989 effectively marked the end of the so-called Cold War between communism and capitalism. The subsequent collapse of the USSR led to large reductions in the nuclear arsenals, and by 1991 the clock had moved back to 11.43. Once again, there were optimistic hopes of an era of peace and an end to the threat of nuclear weapons. It was not to be. The political system in the United States made it almost impossible to scale back arms production.

The Nuclear Non-Proliferation Treaty had been negotiated in the 1970s. It aimed to prevent the spread of weapons beyond the five nations that had already acquired them. But those countries did not implement their promise to disarm, so inevitably other nations decided that they would be more secure if they had nuclear weapons. With the evidence that India, Pakistan and Israel had developed nuclear weapons, the clock moved forward again year by year. It reached 11.53 in 2002. Since then, the managers of the Doomsday Clock have added new threats to the original fear of nuclear war. In 2007 they said, 'climate change also presents a dire challenge to humanity' and advanced the clock to 11.55.

More recent annual reports have warned that international leaders are failing to perform their most important duty – ensuring and preserving the health and vitality of human civilisation. The nations with nuclear weapons continue to test new devices and improved delivery systems. The number of weapons has dropped from its peak of over 70,000 to about 13,000, but that is still enough firepower to wipe out civilisation several times over. And there are new players, including North Korea and perhaps Iran. The 2017 report said: 'It is two and a half minutes to midnight, the Clock is ticking, global danger looms. Wise public officials should act immediately, guiding humanity away from the brink. If they do not, wise citizens must step forward and lead the way.' With no progress towards nuclear disarmament and increasing concern about the equal lack of progress in the effort to slow climate change, the scientists moved the clock forward again in 2019 and left it at a mere 100 seconds to midnight in 2021.

I was interviewed by television and radio programs about the dangers posed by nuclear weapons, but I saw no response from the Australian Government. Admittedly in 2020 and early 2021, even climate change had been swept aside by the more urgent emergency of the COVID-19 pandemic, but it remains a general observation that Australian politicians generally seem quite oblivious to the threat posed by nuclear weapons. The subject is very rarely even discussed in the parliament.

There have been times in our history when Australia has played a constructive role in the global discussions about nuclear weapons, but not in recent years. The Menzies Government had hoped that nuclear weapons would be confined to a small number of countries and Australia supported the NPT in the 1970s. During the final days of the Keating Government, his foreign minister Gareth Evans actively prosecuted the cause of disarmament.[2] The International Court of Justice was asked by the UN General Assembly to rule on whether the threat or use of nuclear weapons would be in breach of international law. The line put forward by Evans was that every nation should be working to eliminate nuclear weapons, but he appeared to give Australia wriggle room by advancing the curious notion of 'stable deterrence'. As Broinowski observes, this was seen as 'code for allowing the United States to keep its nuclear weapons and its bases in Australia'.

As evidence for this view, Broinowski quotes Keating as Prime Minister arguing the so-called joint bases at Pine Gap and Nurrungar, which are actually US bases over which Australia has no control, 'would be important to the verification processes we would have to enter into to achieve a nuclear-weapons-free world'.[3] In a Solomon-like judgement, the court agreed that the threat or use of nuclear weapons was not allowable under international law unless it was in response to a grave threat to national survival. The obvious problem with that argument is that the countries holding nuclear weapons would probably say that they have no intention of using those weapons unless there were a grave threat, but there is no objective definition of what that means.

Keating and Evans also convened a high-powered meeting, the Canberra Commission, bringing together distinguished scholars and experienced defence leaders to discuss ways to rid the world of nuclear weapons.[4] There was a change of government between the first meeting of the commission in early 1996 and their report being produced that August. It called for 'a clear and unequivocal statement' from the nations with nuclear weapons that they intended to work towards elimination, as well as suggesting practical steps towards that goal, including an agreement not to be a first user. That is an interesting concept. If all the nations with nuclear weapons agreed not to be a first user, and they all trusted the others to honour that agreement, there would clearly be no point continuing to hold weapons that will never be used.

Broinowski observes that the incoming Howard Government did nothing to promote the recommendations of the commission.[5] Being even-handed, he also notes that the Keating Government was equally uncritical of US actions that were clearly in conflict with the NPT. Like the notion of 'stable deterrence', it seemed to argue that US weapons were acceptable because they deter other countries from acquiring weapons, whereas other countries having weapons was intrinsically a threat to world peace. When George W. Bush, Tony Blair and John Howard tried to justify invading Iraq, they argued that it was necessary because Iraq had 'weapons of mass destruction'. That was a lie, but critics at the time reasonably asked why it was acceptable for the United States and the United Kingdom to have such weapons, but would have been an affront for a poorer nation to be so armed. The then head of the IAEA, Mohamed ElBaradei, made the reasonable point that as long as some nations based their defence strategy on nuclear weapons, there was no moral argument against other nations following suit.[6] That is a quite fundamental point about nuclear weapons. The threat to use them is essentially a proposal to murder millions of civilians if a nation is unable to get its way by peaceful means. I cannot imagine any moral code that would justify that approach.

In the twenty-first century, things have got steadily worse. At the 2000 UN conference reviewing the NPT, Australia did not support the seven countries that called on the countries holding nuclear weapons to fulfil their obligations under the treaty by taking steps to disarm. Broinowski summarises the Australian approach.[7] Like previous Australian prime ministers, Howard would describe any nuclear disarmament proposal that insisted on the nuclear disarmament of the United States as 'unbalanced'. Equally, an expectation that the United States would agree to a disarmament program that curtailed its right to keep on developing American nuclear weapons and delivery systems would be seen as 'unrealistic'.

After the terrorist attacks on the World Trade Centre on 11 September 2001, US President Bush initiated a Nuclear Posture Review that was leaked to US media in 2002. Broinowski points out that it asserted the United States right to 'threaten the use of nuclear weapons against countries that do not possess them' and scarily listed specific scenarios in which the United States would feel justified to use nuclear weapons such as 'an Arab–Israeli conflict, a Chinese attack on Taiwan, a North Korean attack on South Korea'.[8] In other words, the Bush administration essentially said that it felt the United States had the right to ignore not just the spirit of the NPT but the actual words of the treaty and use nuclear weapons any time it could not achieve its military objectives using conventional weapons.

While subsequent prime ministers have not been as crass as John Howard with his styling of Australia as the USA's deputy sheriff, and few foreign ministers have been as bumbling and sycophantic as Alexander Downer, successive governments have been deafeningly silent about the inevitable impetus the US stance gives to other nations acquiring nuclear weapons. That inevitably increases the risk that some day a leader will be desperate enough or deranged enough to use them.

A UN resolution to outlaw nuclear weapons was circulating in 2021. A 2017 UN conference had developed the treaty and it entered into force early in 2021, having been signed by the required fifty nations.[9] Perhaps understandably, none of the nations with nuclear weapons were among the fifty signatories. Neither was Australia. This was being widely interpreted as an admission that Australia's implicit defence policy is a belief that US nuclear weapons will deter any potential enemy. Since use of nuclear weapons against any country that is also nuclear-armed would inevitably produce retaliation, that policy is based on a hope that the United States would risk annihilation to defend their Australian mates. Leading defence analyst Professor Hugh White has argued that there should at least be a serious conversation about the wisdom of total reliance on the American nuclear shield.[10]

Jim Falk argued that Australia could not rely on the United States in the event of conflict with a regional neighbour.[11] He pointed to the fact that the United States sided with Indonesia and supported the annexation of West Papua 'despite considerable pressure' from the Australia Government. The United States recognised Taiwan as an independent national entity when the Chinese Nationalists fled there after losing the civil war to Mao's communists, but twenty years later, when it became expedient to seek a rapprochement with China, it accepted the traditional Chinese view that Taiwan is a province inseparably linked to the mainland. More recently it seems to have moderated that view and effectively said that it would regard any attempt by China to assert its sovereignty over Taiwan as an act of aggression. When the Falklands War broke out between the United Kingdom and Argentina, the United States declined to provide military support to either side. Falk argued, on the strength of this historical evidence, that the United States would probably not support Australia if we were in a conflict with a neighbouring country. In any case, he reasoned, it was extremely unlikely that any nearby country would ever seek to invade Australia, citing a report to a parliamentary security committee that even the Japanese high command, at the height of their wartime ascendancy in 1942, ruled out the idea of invading Australia as logistically impossible.[12]

In 2021, IPAN's public inquiry was looking at the cost of the Australian willingness to support the United States military in its various adventures, without any serious assessment of whether it was in the country's interest to be involved in those conflicts. There is at least a *prima facie* case that Australia would be better off socially and economically, as well as in security terms, if the nation's leaders made independent assessments of the national interest, rather than mindlessly following wherever clumsy US leaders were going. IPAN drew my attention to startling comments made by Peter Jennings of the Australian Strategic Policy Institute and Sky News host Catherine McGregor about the increasing tension between China and the United States over the future of Taiwan.[13] Brian Toohey points out that Jennings, in a previous role as a political staffer, was one of the people Prime Minister John Howard consulted in 2003 before making his notorious claim that Iraq had weapons of mass destruction.[14]

Jennings described Taiwan as 'the emerging front line of Australia's defence', as if we were obliged to get involved in the increasingly bellicose statements of the two super-powers. McGregor claimed that war between the United States and China is not just a possible outcome of the present posturing but almost inevitable, adding, 'whether we like it or not ... we are integrated into the United States warfighting system ... and due to the location of the US base in Australia, it means we will be "the first target" if China launches an attack against Taiwan. We are in this, there is no option for neutrality so our preparedness in my view is lagging.' This is the sort of talk that was common during the so-called Cold War, when strategic analysts claimed that our perceived role as a US acolyte put us in the front line. The former USSR would not dare attack the United States, they argued, knowing that an all-out nuclear war would be devastating, so they would show their displeasure with American actions by obliterating a couple of Australia cities. That sort of talk strengthens IPAN's argument that Australia would be safer if it were genuinely independent, rather than being seen as a mindless appendage of the United States.

As well as pondering the wisdom of a defence strategy that is predicated on the assumption that US nuclear weapons might protect us, there should also be a public discussion about the wisdom of Australia's uranium export policy. As was foreshadowed when the Fraser and Hawke governments began watering down the strict set of safeguards announced in 1977, there have been several further steps down that slippery slope. Robert Milliken argued in 1980 that whenever the safeguards looked like preventing a commercial deal, it was the safeguards that got jettisoned rather than the deal.[15] The most recent example was the 2012 announcement that the Australian Government would allow export of uranium to India, even though India has not signed the NPT and used Canadian civil nuclear technology to produce nuclear weapons, which in turn inspired Pakistan to follow suit.

The World Nuclear Association (WNA) gives the following figures for the scale of Australian uranium exports, in tonnes of $U_3O_{8,}$, noting that detailed figures have not been available in recent years:

- USA: up to 5000 tonnes per year.
- EU, including Belgium, Finland, France, Germany, Spain, Sweden, UK: up to 3500 tonnes per year.
- Japan: formerly up to 2500 tonnes per year.
- South Korea: up to 1500 tonnes per year.
- China: about 500 tonnes per year.
- Taiwan: up to 500 tonnes per year.
- India: up to 300 tonnes per year likely from 2018.[16]

There is more than a little inflation in the use of the expression 'up to' in most lines of that table. The Department of Foreign Affairs gave the total quantity exported in 2108–19 as 7500 tonnes, the highest figure in recent years. The figures given by the World Nuclear Association add up to nearly double that. The adverb 'formerly' in reference to Japan is an acknowledgement that the shutdown of the entire Japanese nuclear power system after the Fukushima disaster inevitably meant that Japan stopped buying uranium.

In terms of recent export deals, the WNA notes a 2006 agreement was concluded with China, one in 2007 with Russia, one with the United Arab Emirates in 2014, the agreement with India in 2015 and one with Ukraine in 2016. This made the total twenty-five treaties covering twenty-eight countries within the European Union, fifteen other countries and Taiwan, which Australia recognises to be a province of China.

The official Australian Government position is that the following conditions apply to what they call Australian Obligated Nuclear Material (AONM), by which they mean nuclear material subject to an Australian cooperation agreement:

- AONM will be used only for peaceful purposes and will not be diverted to military or explosive purposes
- IAEA safeguards will apply
- Australia's prior consent will be sought for transfers to third parties, enrichment to 20 per cent or more in the isotope U^{235} and reprocessing
- fall-back safeguards or contingency arrangements will apply if for any reason NPT or IAEA safeguards cease to apply in the country concerned
- internationally agreed standards of physical security will be applied to nuclear material in the country concerned
- detailed administrative arrangements will apply between ASNO and its counterpart organisation, setting out the procedures to apply in accounting for AONM

- regular consultations on the operation of the agreement will be undertaken and
- provision will be made for the removal of AONM in the event of a breach of the agreement.[17]

These certainly look like a strong set of conditions that allow governments to say they are confident that Australian uranium is only used for peaceful purposes in nuclear power reactors and research facilities. The abiding concern is, as the Fox Report said, after summarising what it called 'the main limitations and weaknesses of the present safeguards agreements', that 'these defects, taken together, are so serious that existing safeguards may provide only an illusion of protection'.[18] While the countries that buy Australian uranium enter into some form of safeguards agreement, it is difficult to ensure that those undertakings are honoured. As mentioned, when Mohamed ElBaradei was head of the IAEA, he lamented the fact that he was expected to police hundreds of installations around the world with a budget comparable to a metropolitan police force. Broinowski argues that some of our customer countries have, at best, been sloppy in their security, quoting a Far Eastern Economic Review report:

> In May 1994, the Japanese Power Reactor and Nuclear Fuel Development Corporation (PNC) admitted that nearly 70 kilograms of plutonium – enough for several nuclear bombs – was found to be missing during a process in the manufacture of plutonium fuel pellets. PNC later said it had recovered all but 10 kilograms of the material.[19]

It is inevitable that a nuclear reactor using uranium as its fuel will produce significant amounts of plutonium. I remember being told that a standard 1000 megawatt nuclear power station would produce about 200 kilograms of plutonium a year. The fuel elements can be reprocessed to extract the plutonium, either for weapons production or to use as a fuel for other reactors. That is why the safeguards agreements are usually very particular about whether reprocessing can be allowed. Robert Alvarez published a paper in the *Bulletin of the Atomic Scientists* about the problem that arises, even if the fuel rods are not treated.[20] At the end of 2018, he wrote, the US nuclear industry had accumulated spent fuel rods from power reactors containing more than 800 tonnes of plutonium, a scary quantity when about 10 kilograms is enough for a bomb. The good news, Alvarez wrote, is that 'the intense radiation of used nuclear fuel assemblies makes them essentially impervious to theft or diversion to weapons use'. The concern is that 'a great deal of the radiation barrier protecting them will have decayed' in 300 years. That is clearly an issue for the long term.

As well as the risk of uranium being used to produce nuclear weapons, there are the abiding concerns about safety and waste management. A 2021 article by Jim Green and David Noonan was published in *The Ecologist*, arguing that Australian uranium exporters should accept their partial responsibility for the Fukushima disaster. The Australian Nuclear Safeguards Office has acknowledged that AONM was in at least five of the six Fukushima reactors and possibly in all of them. Green and Noonan argue:

> The Nuclear Accident Independent Investigation Commission – established by the Japanese Parliament – concluded in its 2012 report that the accident was 'a profoundly man-made disaster that could and should have been foreseen and prevented' if not for 'a multitude of errors and wilful negligence that left the Fukushima plant unprepared for the events of March 11'. The accident was the result of 'collusion between the government, the regulators and TEPCO', the commission found.[21]

In a sense, there was nothing new in this revelation. Broinowski wrote that Japanese faith in the regulation of its nuclear industry had been shaken by 2002 revelations:

> The Tokyo Electric Power Company (TEPCO), one of Australia's main customers for uranium, had for sixteen years covered up numerous cases of damage at three nuclear power plants during the 1980s and 1990s. It had also falsified inspection reports in an attempt to hide the fact that previous safety reports were incorrect.[22]

Green and Noonan argue that the Australian uranium industry did not react to those shocking accounts of Japanese mismanagement of nuclear reactors, nor to the 2007 reports of hundreds of incidents of malpractice or growing internal criticism of the vulnerability of reactors to earthquakes or tsunamis.[23] Even after the Fukushima disaster, they write, Japanese authorities were eager to find ways to get the nuclear industry going again, quoting the committee that investigated the accident: 'Unfortunately, the new regulatory regime is … inadequate to ensure the safety of Japan's nuclear power facilities. The first problem is that the new safety standards on which the screening and inspection of facilities are to be based are simply too lax.'

While it is true that the new rules are based on international standards, the international standards themselves are predicated on the status quo. They have been set so as to be attainable by most of the reactors already in operation. In essence, the NRA made sure that all Japan's existing reactors would be able to meet the new standards with the help of affordable piecemeal modifications.

The critique raised again the whole issue of product stewardship and the uranium industry's responsibility for the product it exports. It inevitably reminded me of the exporters of coal and gas from Australia, who use what has been called the drug-dealer's defence to excuse themselves from the charge of contributing to climate change. They say that they don't burn the coal or gas, they just sell it, so the charge of accelerating climate change should be levelled at the power companies who burn the fossil fuels rather than the Australian exporters. Like drug-lords, they also often say that if they weren't providing the product, somebody else would, perhaps people with even lower standards than them who would sell a dirtier product. Of course, the legal system prosecutes drug dealers because it holds them responsible for their actions. I believe that it is entirely reasonable to apply the principle of product stewardship and say that mineral exporters are accountable for the uses of their product.

Just as the Fox Report said that 'the nuclear power industry is unintentionally contributing to the risk of nuclear war', it is obvious that the fossil fuel industry is *intentionally* contributing to the risk of accelerating climate change. It would be a responsible acceptance of our international obligations to have a policy of phasing out fossil fuel exports. Instead, Commonwealth and state governments have gone to considerable trouble to expand gas exports and even support the opening of whole new coal export provinces such as the Galilee Basin. Future generations will find it difficult to believe that elected politicians were still behaving with such crass disregard for their responsibilities in 2021.

At one level, the community accepted that principle of product stewardship when it endorsed the imposition of safeguards agreements as a pre-requisite of uranium exports. The willingness of politicians to accept continuing erosion of those principles is another example of the tendency to see economic considerations as trumping all others.

Activities will be tolerated despite environmental damage or social disruption or increasing inequity or even increasing the risk of nuclear devastation as long as they are contributing to economic growth. Broinowski's depressing conclusion to his wonderful book about Australia's role in the nuclear industry is that politicians have, for fifty years, 'handled one of the most dangerous of known substances in a way that shows they care much less for national or international security interests than they do for their own political survival'.[24]

Two of the recommendations of the Fox Report are worth repeating:

> Policy respecting Australian uranium exports, for the time being at least, should be based on a full recognition of the hazards, dangers and problems of and associated with the production of nuclear energy, and should therefore seek to limit or restrict expansion of that production.

> Policy with regard to the export of uranium should be the subject of regular review.[25]

It is also worth recalling one of the conclusions of the UK Royal Commission, quoted in the Fox Report:

> There should be no commitment to a large programme of nuclear fission power until it has been demonstrated beyond reasonable doubt that a method exists to ensure the safe containment of long-lived, highly radioactive waste for the indefinite future.[26]

It is now nearly fifty years since the discovery of the large uranium deposits in the Northern Territory began public discussion of the wisdom of having a large-scale uranium export industry. Rather than recognising 'the hazards, dangers and problems of an associated with the production of nuclear energy', successive governments, state, territory and Commonwealth, Coalition and ALP, have been prepared to overlook those in pursuit of short-term economic gains. There is, globally, a large program of nuclear fission power and there has been a concerted effort to expand that activity with the dubious claim that it could help to avoid the climate emergency we face. Rather than the sort of regular review that the Fox Report called for, successive governments have pandered to the short-sighted ambitions of commercial operators and wilfully ignored the risks associated with expanding exports.

It is surely time to take stock and weigh up whether the relatively trivial economic outcome, less than Australia makes selling cheese, is worth the increasing risks.

CONCLUSION

My lifetime has seen a gradual erosion of trust in experts. When I was young, we accepted that scientists, engineers and medical practitioners spoke with authority. GPs tell me that patients now often arrive at a surgery with their own diagnosis and even recommended treatment, having consulted Dr Google about what they believe to be the relevant symptoms. I mentioned earlier an Australia prime minister who said that he preferred to trust his instincts about climate change rather than listen to what the scientific advisors were saying. Another recent prime minister told the audience at a political rally that climate science was 'crap', while the prime minister at the time of writing this book had waved a lump of coal around in the parliament to demonstrate his willingness to ignore the scientific advice. A member of parliament elected to represent the One Nation Party has even made speeches claiming that the climate isn't changing, accusing thousands of scientists of being in a huge global conspiracy to falsify the temperature record.

I have a suspicion that the erosion of trust is at least partly a con-sequence of the indefensible assertions made about nuclear energy by some of its proponents. In overall terms, they assured the public that nuclear power was cheap, clean and safe. I have given in earlier chapters examples of the sorts of claims being made in the 1970s. The cover-up of the Windscale fire by the UK Government was an attempt to main-tain the image of safety, but the assurances were rocked by the Three Mile Island meltdown. At least the public could be told that there had been no risk to the surrounding community. The Chernobyl disaster was much more serious, but Western communities were promised that it was a consequence of clumsy Russian technology and the lack of a safety culture. Fukushima was the final blow, raising obvious concerns that nuclear reactors might just be too dangerous if a sophisticated modern nation like Japan could suffer such a catastrophic failure.

The proposition that nuclear power stations are relatively clean environmentally had more substance. It is true that they contribute less to global climate change than power stations burning fossil fuels. They even release less radiation when working normally than a large coal-fired generator, because coal usually contains small amounts of uranium, and a nuclear power station does not produce anything like the air pollution from burning coal. The substantial environmental problem, however, is radioactive waste. Most of the forty nations using nuclear power are just piling up the spent fuel, with no clear plan to manage it for the incredibly long periods required. That is a very serious environmental legacy for future generations.

As for nuclear power being cheap, it was true that some of the earliest nuclear power stations were reliable and ran for thirty to forty years, giving relatively inexpensive electricity. The modifications required more recently to assure the public of the safety of nuclear reactors, however, have dramatically inflated the delivered cost of power, now beyond the reach of even the most creative arithmetic. By contrast, large-scale renewables systems such as solar farms and arrays of wind turbines are still coming down in price, leading the electricity industry globally to have almost abandoned nuclear power.

Problematically, scientists and engineers working in the nuclear industry tend to believe they are objective experts, relying on facts. The 1977 UK inquiry into a proposed reprocessing facility for nuclear fuel became, as the Fox inquiry had, a broader examination of the whole nuclear industry. One of the industry's senior managers lamented in the house journal *Atom* what he saw as 'the ability of objectors to make emotional and unquantified statements', whereas scientists and engineers 'are constrained by the need for accuracy'.[1] The *Guardian's* analysis of the evidence concluded that the fears were not irrational, noting that one engineer conceded that 'a generation's experience' was no guarantee of perpetual safety.[2] Brian Wynne saw the nuclear industry's contribution to the inquiry as 'a ceremonial of collective self-delusion' from an elite 'deeply immersed in their own myths'.[3]

To be fair, nuclear engineers are not the only experts whose assurances have proved false and eroded public confidence. When there was an outbreak in the United Kingdom of 'mad cow disease', bovine spongiform encephalopathy, the prime minister of the time persuaded some scientists to make public statements assuring the community that it was safe to eat beef because the disease could not be transmitted to humans. The subsequent outbreak of variant Creutzfeldt-Jakob disease was linked to beef consumption. It was a devastating blow to public confidence in expert advice.[4]

There is also the problem of conflicting expert advice. There are legitimate differing views about such questions as whether a new generation of nuclear reactors can be developed to avoid the problems that have emerged. As well as legitimate differences, the last few decades have seen the spread of what could be called illegitimate differences, acts of intellectual bastardry to manufacture public doubt and reduce the impetus for action. Naomi Oreskes and Erik Conway document in their book the way the tobacco industry worked to spread doubt about the potentially lethal consequences of smoking, allowing them to continue not just selling but vigorously promoting a product that shortens the lives of half its users.[5] When the science of climate change became clear about twenty-five years ago, the fossil fuel industry had a choice. It could have accepted the science and worked to revise its practices, embracing the new cleaner technologies. Instead, it chose to fund the spreading of misinformation to assist those politicians who wanted to delay taking action. In Australia, we have seen the recruitment of qualified scientists with no expertise in climate science or the underlying physics to spread doubt about the scale of the problem and the urgency it demands. Future generations will live with more severe climate change than would have occurred if concerted action had been taken when the warnings from the science were clear.

The problem for decision-makers is how they should respond when the experts don't speak with one clear voice and they are giving conflicting advice. I remember being on a panel in 1988 with the then Minister for Science, the Honourable Dr Barry Jones. A formidable intellect, Dr Jones is the only person to have been elected a Fellow of all four of the Australian learned Academies. In 1988, there was still significant disagreement about climate change. It was accepted that human activity was changing the levels of greenhouse gases in the atmosphere and basic physics led to the conclusion that this would alter the climate, making the Earth warmer. It was also accepted that warming was happening as well as other changes, like alterations of rainfall patterns and increasing sea levels. However, most cautious scientists were saying that it was not proven that the climate changes being observed were being *caused* by the increasing quantities of greenhouse gases. I thought the link was clear enough to write a paperback book, *Living in the Greenhouse*, in 1989, but at the time that was a minority view in the scientific community. The United Nations established the Intergovernmental Panel on Climate Change (IPCC) to review the evidence and by 1995 the debate was essentially over, but in 1988 there were still legitimate differences.

When Dr Jones was asked how decision-makers should respond, he gave characteristically wise advice. He said we should always consider the consequences of being wrong. If those scientists who believed that climate change was a serious problem being caused by human activity were wrong but we took their advice, he said, the worst that would happen would be large-scale use of cleaner but more expensive energy. That might not be economically optimal, he said, but it would not be a serious risk. On the other hand, he said, if those scientists are right and we don't listen to them there could be disastrous outcomes, even potentially the collapse of civilisation. So, he said, even if I thought there was only a 5 per cent chance they were right, I would be urging us to respond.

A parallel argument could be applied to nuclear power and nuclear weapons. We should consider the consequences of being wrong. It may eventually be possible to develop nuclear reactors that are as safe as the Rasmussen study predicted, with only one serious accident in the next million years, while also devising systems for the safe long-term management of the radioactive waste from nuclear power stations, but the consequences of failure in either area are too awful to contemplate. Large areas around Chernobyl and Fukushima remain unsafe for humans, while there is not yet a proven multi-barrier system to provide public assurance of the safety of waste management. Similarly, it might turn out that the world can live under the shadow of nuclear weapons, that my fear the spread of weapons beyond the 13,000 or so currently held by nine countries will lead to them being used by a desperate or deranged leader is misplaced, but again the consequences of being wrong are appalling.

I saw an article published in a UK paper by a staff correspondent Beau Bilinovich. After discussing instances when only a stroke of luck prevented catastrophic nuclear war, he concluded:

We are left with no other option than to confront the truth.

Those entrusted with the authority to deploy and launch these missiles at a moment's notice cannot be trusted. The systems designed to monitor attacks cannot be trusted. Foreign nations in possession of this same deadly tool cannot be trusted. While we may think we can handle nuclear weapons, reality shows the opposite. In truth, no one can be trusted with nuclear weapons. If we do not realize this, we may not have any more stories to tell.

Our inability to trust anyone with these weapons demands that we abolish them. The sooner we accomplish this goal, the safer the world becomes. Getting rid of these weapons is the only way to avoid a nuclear apocalypse.[6]

Of course, the existence of nuclear weapons does not guarantee that they will be used and end civilisation, but the consequences of being wrong about that are so appalling as to demand concerted action. Toohey documents the frightening mishaps that could have caused global nuclear war: a defective computer chip, a training tape being mistaken for a real attack, 'false alarms triggered by the moon rising over Norway, the launch of a weather rocket, a solar storm, sunlight reflecting off high-altitude clouds and a faulty telephone switch in Colorado'.[7] In both the USA and the former USSR, individuals who bravely ignored their training and the signals they were receiving saved the world from the horror of thermonuclear war.

I recently came across a column on the subject, actually published by the Murdoch press nearly ten years ago, which summed up beautifully the case for Australia to act responsibly for nuclear disarmament:

We have all the elements of a perfect storm for nuclear calamity – ample supply of nuclear materials usable for weapons and the means to make them, inadequate nuclear security arrangements in many countries, easy access to the knowledge to build a bomb on the internet and groups proclaiming malicious intent to acquire and use a nuclear weapon.

The risks of nuclear escalation of a conflict between India and Pakistan, in the Middle East or on the Korean peninsula are real and present dangers. And thousands of nuclear weapons from bloated cold war arsenals persist on high alert, prone to things going wrong in a crisis, technical failure, error and cyber attack.

We are at a crossroads: during the next several years either we will take decisive steps towards banning nuclear weapons or more governments will become nuclear capable and the risks of nuclear weapons being used will grow towards their inevitable use.

With over 30 per cent of the world's uranium, Australia has both an opportunity and a responsibility to become a global leader in galvanising international cooperation towards a nuclear weapons free world …

The column criticised Australia's silence at the UN when other countries called for an immediate start to negotiations to rid the world of nuclear weapons, as well as moves to allow uranium sales to countries 'with patchy security arrangements'. It concluded:

Let us not wait until a time when the unthinkable happens and a nuclear weapon is detonated on one of the world's great cities, and we wonder what we could have done to halt the nightmare that would unfold before our eyes. It is time for Australia to adopt a nuclear-weapon-free defence posture and join other nations in working to achieve a comprehensive, verifiable treaty to abolish nuclear weapons.

You are probably wondering what wild-eyed radical or naive idealist might have written those words. The author was the former Liberal Party leader and Australian prime minister from 1975 to 1983, Malcolm Fraser.[8] The architect of the decision to allow the mining and export of uranium from the Ranger mine, he was thirty-five years later with hindsight calling for a more responsible approach.

The export of Australian uranium is certainly producing high-level radioactive waste for which there is currently no proven safe permanent disposal. It also puts into circulation fissile material that can be used to manufacture nuclear weapons. We are morally responsible for those consequences. For an export revenue comparable with minor minerals like tin and tantalum, less than we get from exporting cheese, I believe the risks are unacceptable.

The purpose of writing this book, was to draw attention to the consequences of Australian enthusiasm for mining and exporting uranium, hopefully to catalyse the review of the policy the Fox Report said we should have regularly. Our decisions are creating the future in which our descendants will live. I believe we should be working to give them the legacy of a clean and secure future, not one clouded by radioactive waste and fear of nuclear war.

I remember watching a film about the US nuclear weapons tests in the Marshall Islands. The radioactive fallout caused dreadful problems for the island communities, both in direct health impacts and in their damage to the marine ecosystems. In a closing scene, an Islander woman asked how 'clever people' could let these things happen. After reflecting, she concluded that the nuclear scientists were 'clever at doing stupid things'. History will probably see that as an appropriate overall comment on the nuclear industry. It has been clever at doing stupid things, creating appalling problems for our descendants. We have a responsibility to phase out our contribution to those problems.

APPENDIX:
ARPANSA AND ITS
ADVISORY BODIES

The regulator of all aspects of radiation is the Australian Radiation Protection and Nuclear Safety Agency (ARPANSA). It was established by an Act of the Commonwealth Parliament in 1998, replacing two earlier bodies. It is responsible for licensing the research reactor at Lucas Heights and will be required to consider providing a licence if a proposal comes forward for a national radioactive waste storage facility. The public hearing in Adelaide was part of the process when ARPANSA was considering whether to approve an earlier proposal for a low-level waste repository near Woomera. That proposal was withdrawn after the public hearings and the SA State Government subsequently legislated to strengthen earlier legislation prohibiting the storage of radioactive waste from other states and territories.

The ARPANSA Act sets out the responsibilities of the regulator. They are to:

- Identify, assess and communicate health, safety and environmental risks from radiation
- Promote radiological and nuclear safety and security, and emergency preparedness
- Promote the safe and effective use of ionising radiation in medicine
- Ensure risk-informed and effective regulation
- Enhance engagement with stakeholders.

In carrying out its work, ARPANSA is advised by three representative bodies: the Radiation Health and Safety Advisory Council (RHSAC), the Nuclear Safety Committee and the Radiation Health Committee. The RHSAC includes the CEO of ARPANSA and several external members, including one who represents the interests of the general public. Membership includes two radiation control officers from different states, a representative of the Northern Territory Government and seven members appointed for their technical expertise, as the website says, in various aspects of radiation protection 'including medicine, mining, regulation, health research and law'.

The member representing the interests of the general public in 2021 was Dr Peter Karamoskos, a consultant radiologist at Epworth Medical Imaging in Victoria. Members are appointed by the Federal assistant minister for health for terms of three years. The RHSAC advises the CEO on emerging issues and matters of major public concern relating to radiation protection and nuclear safety. Among other functions, it develops codes and standards on the advice of the two committees listed above. Those bodies are appointed by the CEO to provide specialist advice. The Radiation Health Committee advises the CEO and the Council on matters relating to radiation protection, including the formulation of draft national policies, codes and standards, while the Nuclear Safety Committee, as its name implies, advises the CEO and the council on matters relating to nuclear safety, including the monitoring of practices and procedures.

I was a member of the Council from 2004 to 2016. As the designated member representing the interests of the general public, I was the point of contact for people concerned about various aspects of radiation protection, ranging from tanning studios to uranium mines and nuclear medicine. Because it is an issue related to electromagnetic radiation, I even received a submission from a group of parents who were worried that Wi-Fi in their children's school could be a health risk. It was also my function to provide a critical review when codes and standards were being developed, after discussion with the relevant industry such as uranium mining or nuclear medicine.

I was acutely conscious of the problems that had occurred in other countries when there was a cosy relationship between the regulator and industry with little concern for the public interest. I took very seriously my responsibility to ensure that the public interest was taken into account in deciding what would be appropriate standards. As a specific example, in the case of medical imaging, the council went to a great deal of trouble to work out with experts the minimum radiation dose to provide accurate diagnosis. It is, of course, important to ensure that diagnostic imaging does more good than harm. When I was young, tuberculosis was a serious health problem in Australia. Most cities had specialised facilities for treating those who suffered from the disease. In New South Wales where I lived there were two, both well inland from Sydney – one in the Blue Mountains and one in the Southern Highlands. It was believed that the mountain air was good for those with the disease. At that time, all adults were required to have regular chest X-rays to identify those having tuberculosis. The practice was later discontinued when so few cases were being detected that health authorities concluded the imaging was probably doing more harm than good.

In similar terms, I remember more recently ARPANSA, acting on the discussion at the advisory council, discouraging the use of whole-body CT scans in the absence of evidence of an established health condition, especially for children.[1] Its advice has also led to reduced doses being used for medical imaging.[2] Charles Pope estimates that a lumbar CT scan increases the risk of dying from cancer from 20 per cent to 20.05 per cent.[3] That is not a huge risk, but it is also not negligible, confirming the argument that such a procedure should be undertaken only when there is clear evidence it will help to treat a serious problem.

The ARPANS Act specifically provides that the regulator is independent of government. One feature of its work, which should give the community confidence in the work of the organisation, is the close links it has with international best practice. The current CEO, Carl-Magnus Larsson, came to ARPANSA after an impressive career in the Swedish regulatory body and experience in such international bodies as the UN Scientific Committee on the Effects of Atomic Radiation, UNSCEAR. In that capacity, he was centrally involved in the international investigation of the Fukushima disaster. He is also vice-president of the Conference of Parties to the Convention on Nuclear Safety. His deputy CEO, Dr Gillian Hirst, was recently appointed chair of UNSCEAR.

As the minutes of Council meetings are publicly available on the ARPANSA website, I noticed that the current member representing the public interests has received a communication from the Barngarla group concerned about the proposed national low-level waste facility. The Council correctly advised the group that the regulator has no role in selecting a site for such a facility. ARPANSA sets standards that would have to be adopted for a facility to be established and if a formal proposal is developed, the regulator will then decide whether to issue a licence to allow the proposed activity. The regulator is free to decide what would be the appropriate process for considering a project. When there was the proposal for a national disposal site for low-level radioactive waste near Woomera, ARPANSA conducted public hearings in South Australia. Then CEO Dr John Loy asked me, the member of the advisory council representing the public interest, and an overseas expert on radioactive waste storage, to assist him in hearing submissions and preparing a report on the proposal.

As the regulator of all aspects of radiation, ARPANSA is responsible for ensuring that the mining and milling of uranium does not cause environmental damage or risk the health of workers.[4] When I was on the advisory council, I visited Roxby Downs and was assured of the measures taken to protect miners. Now that the Ranger mine is winding up, ARPANSA will be required to oversee the closing down procedures and rehabilitation of the site. Energy Resources Australia (ERA) announced in January 2021 that production would cease, noting both the current state of the world uranium market and the fact that the company had failed to receive approval of native title holders for continuation of the mining operation. Production had obviously been slowing, with 390 tonnes exported in the December quarter of 2020 and about 1600 tonnes in the entire year, compared with average figures of about 3000 tonnes a year over the life of the mine. The 2021 announcement noted that ERA 'has previously faced several cost blowouts in rehabilitating the site'.[5] Tailings are being transferred from a temporary storage site to pit three of the mine and revegetation of pit one is continuing. The work of rehabilitation is expected to take until 2026. The company has stated that its mine closure plan will deliver a 'positive legacy', but it is difficult to see how it could possibly leave the site in better condition than it was found fifty years earlier. The absolute best that could be achieved would be to leave it in no worse condition, while even that is usually an aspiration of mine site rehabilitation rather than an achievement. It would be lovely if one day there might be no perceptible boundary between the Kakadu National Park and the former mine site, but I am not holding my breath.

Broinowski is critical of the work done by regulators over the decades of uranium mining, lamenting what he sees as public indifference to the fact that substantial areas of Australia, including artesian water resources, have been contaminated.[6] He contrasts the lack of concern about that contamination with the continuing opposition to the establishment of facilities for storing radioactive waste. It is a fair comment. He also criticised the licensing of the OPAL reactor without there being a clearly defined plan for managing the high-level waste that will result, as discussed earlier. In 2018, the ABC reported that ten years of putting spent fuel rods into a water storage the size of a small swimming pool had brought the facility to its capacity, so the spent fuel was being shipped to France for reprocessing. Under the agreement with the French nuclear authority, which operates the plant at La Hague, the waste will have been returned to Lucas Heights, as was done with the previously reprocessed waste for the earlier reactor. It will be there in a temporary store, awaiting a decision about its final disposal. The NSW Government is uneasy about the waste being held at Lucas Heights, but several local authorities are also reluctant to have nuclear waste transported through their road systems, producing an obvious dilemma for ANSTO.

All the recent statements by the Radiation Health and Safety Advisory Council are available on the ARPANSA website. Among those which had been recently released when this book was being written were statements about the Linear Non-Threshold model of impacts of ionising radiation, the risks associated with transport of radioactive material, advice on the role of ARPANSA in dealing with nuclear or radiological emergencies and the problem of naturally occurring radioactive materials, usually referred to by the acronym NORM. This last issue is one that the regulator and its advisory bodies have been grappling with for several years. The recent statement concludes that ARPANSA's two documents, RPS 15 *Guide on Management of Naturally Occurring Radioactive Material* and RPS 9 *Code of Practice for Radiation Protection and Radioactive Waste Management in Mining and Mineral Processing*, have proved to be excellent practical documents for the management of NORM. The Council sensibly supports a graded approach, since the level of concern is related to the degree of radioactivity, and urges a uniform national approach rather than separate regulations in each jurisdiction. Again, that makes sense because there is absolutely no reason to believe that Victorians are affected by radiation differently to South Australians or Queenslanders.

The issue of the safety of transporting radioactive substances has been of public concern for many years. I still have papers relating to discussion of the issue in 2011 when the CEO of ARPANSA invited the Council 'to address the perception that the transport of radioactive materials is more dangerous than the transport of other dangerous goods'. It will be an issue of importance if a national waste facility is established, because the existence of a centralised waste storage must mean that waste will be transported to it from a very large number of sites around the country. Statistically, the experience from other countries is reassuring and the record in Australia is very good. In a typical year, about 5000 tonnes of yellowcake leaves Australia and close to 10,000 orders of medical radioisotopes are delivered, without there being any recorded incidents exposing the community to radiation. The problem is that we do not have any agreed standard about acceptable levels of risk and perceptions often bear no resemblance to the calculated risk.

I remember talking to students about the obvious risks associated with driving a car, a danger that is so predictable we actually talk about 'the road toll', as if the deaths and injuries were an inevitable consequence of using the roads. At one point I calculated that an individual living their entire adult life in Queensland had a 2 per cent chance of being killed and a 20 per cent chance of being injured in a road accident. The students were shocked, but most had a reason for believing that the statistical risk did not apply to them. Those who knew they were not unusually careful tended to believe they were more skilful than the average driver, whereas those who knew they were not unusually skilful generally believed they were extra careful. There is a parallel with perceptions of the risk of transporting radioactive materials. The belief that they are unusually dangerous results in restrictions on where they can be moved, increasing the distances they are carried and therefore increasing the risk of transporting them. Prohibiting export of yellowcake from a particular port will mean it is carried longer distances to another wharf, increasing the risk of a road accident. The obvious conclusion is that decisions about transport of radioactive materials, as for any hazardous substances, should be based on an overall risk assessment. Whether the materials being transported are small medical isotopes, large volumes of uranium oxide or highly radioactive spent nuclear fuel, the most important consideration is packaging to reduce the chance of exposure in the unlikely event of an accident.

While there is always room for improvement, I think the Australian community has been well served by ARPANSA as the independent regulator of nuclear and radiation safety. I am clearly not an objective observer, having represented the public interest on the advisory council for several years, but I believe that position is a very useful safeguard against the sort of regulatory capture that has led to ineffective policing of nuclear activities in some countries. The recent assessments of the situation in Japan are a dreadful warning of what can go wrong if regulation becomes ineffective.

NOTES

Introduction

1 Hughes, Does Australia need Nuclear-Powered Submarines and a Nuclear-Power Sector?
2 Cockburn and Ellyard, *Oliphant*, pp. 95–126.
3 Martin, *Nuclear Knights*, p. 23, p. 45.
4 ibid., p. 31, p. 47, p. 49.
5 Fox et al., *Ranger Uranium Environmental Inquiry First Report*.
6 Switkowski, Uranium Mining, Processing and Nuclear Energy.
7 Scarce, *Nuclear Fuel Cycle Royal Commission Report*.
8 Lowe, *Reaction Time*.
9 Brook and Lowe, *Why vs Why*.
10 Weinberg, 'Science and Trans-Science'.
11 Fox et al., op. cit., pp. 5–6.
12 Collingridge, *The Social Control of Technology*, p. 124.
13 Fox et al., op. cit. p. 6.
14 Douglas et al., *W(h)ither Australia?*

Chapter 1: The Dawn of the Nuclear Age

1 Cawte, *Atomic Australia*, p. 3.
2 Melbourne, Australia's Radium Legacy Waste, pp. 4–15.
3 ibid., pp. 16–18.
4 Pope, *Living with Radiation*, pp. 131–32.
5 Ibid., pp. 2–3.
6 Cockburn and Ellyard, *Oliphant*.
7 Cawte, op. cit., p. 2; Broinowski, *Fact or Fission?*, pp. 15–16.
8 Hore-Lacy and Hubery, *Nuclear Electricity*, p. 64.
9 Weinberg, 'Science and Trans-Science'.
10 International Commission on Radiological Protection, 2007 Recommendations ...
11 Bertell, *No Immediate Danger*, pp. 42–45.
12 Australian Radiation Protections and Nuclear Safety Agency (ARPANSA), *Recommendations for Limiting Exposure ...*
13 Pope, op. cit., p. 144.

Chapter 2: From Basic Physics to Awesome Weapons

1 Broinowski, *Fact or Fission?*, pp, 15–19; Cockburn and Ellyard, *Oliphant*, pp. 110–26.
2 Collingridge, *The Social Control of Technology*, pp. 130–31.
3 Moss, *Men Who Play God*.

Chapter 3: Australia and the British Bombs

1 Broinowski, *Fact or Fission?*, p. 16.
2 ibid., p. 17.

3   Cawte, *Atomic Australia*, p. 4.
4   ibid., pp. 97, 101; Broinowski, op. cit., p. 20.
5   Cawte, op, cit., p. 81.
6   ibid., p. 70.
7   Pope, *Living with Radiation*, pp. 139–40.
8   Cawte, op. cit., p. 72.
9   Milliken, *No Conceivable Injury*, p. 56.
10 Broinowski, op. cit., pp. 30–31.
11 Cawte, op. cit., p. 59.
12 Broinowski, op. cit., p. 31; Milliken, op. cit.
13 Cawte, op. cit., p. 63.
14 Martin, *Nuclear Knights*.
15 Milliken, op. cit., p. 340.
16 ibid., p. 341.
17 ibid., p. 329.

Chapter 4: The Australian Atomic Energy Commission

1   Cawte, *Atomic Australia*, p. 14.
2   Cockburn and Ellyard, *Oliphant*, p. 149.
3   Cawte, op. cit., p. 18.
4   ibid., p. 22–33.
5   Broinowski, *Fact or Fission?*, pp. 32–33; Cawte, op. cit., pp. 44–45.
6   Martin, *Nuclear Knights*, p. 42.
7   Cawte, op. cit., p. 105.
8   ibid.
9   Green, The push for nuclear weapons in Australia 1950s–1970s.
10 Collingridge, *The Social Control of Technology*, p. 126.
11 Findlay, *The Future of Nuclear Energy to 2030 and its Implications for Safety, Security and Non-proliferation*, p. 30.
12 Moyal, 'The Australian Atomic Energy Commission'.
13 Broinowski, op. cit., pp. 42–49.
14 Martin, op. cit., p. 45.
15 Broinowski, op. cit., p. 56.
16 Pope, *Living with Radiation*, p. 47.
17 Cawte, op. cit., p. 168.
18 Broinowski, op. cit., pp. 54–55.
19 ibid., p. 55.
20 ibid., p. 56.
21 ibid.
22 Cawte, op. cit., p. 131; Broinowski, op. cit., p. 68.
23 Cawte, op. cit., p. 132; Broinowski, op .cit., p. 69.
24 Martin, op. cit., p. 47.
25 ibid., p. 30.
26 Broinowski, op. cit., pp. 69–70.
27 ibid., p. 71.
28 ibid., p. 72.
29 Australian Science, Technology and Engineering Council (ASTEC), *Nuclear Science and Technology in Australia*.

30 Australian Nuclear Science and Technology Organisation (ANSTO), The Case for a New Reactor.

Chapter 5: Ranger, the Fox Report and Uranium Exports

1 Fox et al., *Ranger Uranium Environmental Inquiry First Report*, p. 9.
2 ibid., p. 10.
3 ibid., pp. 5–6.
4 ibid., p. 6.
5 ibid.
6 Martin, *Nuclear Knights*, p. 58.
7 ibid.
8 Fox et al., op. cit., p. 6.
9 ibid., p. 185.
10 Cawte, *Atomic Australia*, p. 156.
11 ibid., p. 157.
12 Fraser, Answer to Questions.
13 Alvarez et. al.; Report, pp. 80–91.
14 Fox et al., op. cit., pp. 25–30.
15 Pollitt, 'UK final electricity demand by sector 1960–2009'.
16 Chapman, *Fuel's Paradise*, pp. 97–109.
17 Abdulkarim and Lucas, 'Economies of scale in electricity generation in the United Kingdom'.
18 Fox et al., op. cit., p. 44.
19 ibid., pp. 54–56.
20 Hoyle, *Energy or Extinction?*
21 Hore-Lacy & Hubery, *Nuclear Electricity*.
22 ibid., p. 32.
23 Fox et. al., pp. 82–83.
24 ibid., pp. 185–86.
25 ibid., pp. 187–88.

Chapter 6: The Politics of Uranium in the 1970s and 1980s

1 Byrne, 'Wilson's White Heat'.
2 Broinowski, *Fact or Fission?*, p.39.
3 ibid., p. 108.
4 ibid., p. 109.
5 ibid., pp. 119–125.
6 ibid. p. 125.
7 Amalgamated Metal Workers & Shipwrights Union et al., *The Uranium Debate*.
8 Cawte, *Atomic Australia*, p. 156.
9 ibid., p. 155.
10 Broinowski, op. cit., p. 135.
11 ibid., p. 139.
12 ibid., p. 141.
13 ibid., p. 142.
14 ibid., p. 143.
15 Cawte, op. cit. pp. 157–58.
16 Milliken, 'To Russia With Love'.

17   Broinowski, op. cit. p. 147.
18   ibid., pp. 147–49.
19   ibid., pp. 154–58.
20   Cawte, op. cit. pp. 167–68.
21   Middleton, 'No Uranium Mining in SA'.
22   ibid., p. 166.
23   Select Panel of the Public Enquiry into Uranium, 'The Report of the Public Enquiry into Uranium Mining and Milling Australia'.
24   Australian Science, Technology and Engineering Council (ASTEC), *Australia's Role in the Nuclear Fuel Cycle*.
25   Lowe, 'Blinded by Science'.
26   Suter, *Australia and the Nuclear Choice*.
27   Broinowski, op. cit., pp. 169–70.
28   Lowe, op. cit., pp. 140–41.
29   *The Australian*, 'Editorial'.
30   ASTEC, op. cit., p. 5.
31   Ringwood et al., 'Immobilisation of high level nuclear reactor wastes in SYNROC'.
32   White et al., 'Radwaste Immobilization by Structural Modification'.
33   Broinowski, op. cit., pp. 176–78.
34   Broinowski, op. cit., pp. 177–78.
35   Christoff, 'The Nuclear Disarmament Party'.
36   ibid.
37   Watters and Chandra, *The Nuclear Power Industry*.

Chapter 7: Chernobyl, Climate Change and Fukushima

1   Penney et al, Report on the accident at Windscale No. 1 Pile on 10 October 1957.
2   Leatherdale, 'Windscale Piles'.
3   U.S. Nuclear Regulatory Commission, Reactor Safety Study WASH-1400.
4   Morone & Woodhouse, *The Demise of Nuclear Energy?*, pp. 85–103.
5   Collingridge, *Technology in the Policy Process*, p. 46.
6   Pollard, *The Nugget File*, p. 3.
7   ibid., p. 75.
8   Haynes and Bojcun, *The Chernobyl Disaster*.
9   World Nuclear Association, *Chernobyl Accident 1986*.
10   Morone and Woodhouse, op. cit., p. 100.
11   ibid., p. 101.
12   Myhra, 'Three Mile Island and Implications for Australian Uranium Mining'.
13   Macintyre, 'How the miners' strike of 1984–85 changed Britain for ever'.
14   Nathan, 'Energy and Thatcher'.
15   AP News, 'Hot weather forces 4 French nuclear reactors to shut down'.
16   ElBaradei, 'Director General's Statement to 2005 Review Conference of the Treaty on Non-Proliferation of Nuclear Weapons'.
17   Sabbagh, 'Cap on Trident nuclear warhead stockpile to rise by more than 40%'.
18   Hindmarsh, *Nuclear Disaster at Fukushima Daiichi*.
19   ibid., p. 3.

20 Yamamitsu, 'Ten years on, Japan mourns victims of earthquake and Fukushima disaster'.
21 Murakami and Sheldrick, 'Climbing without a map'.
22 World Nuclear Association, *Fukushima Daiichi Accident*.
23 Hara, 'Social Structure and Nuclear Power Sitting Problems Revealed', p. 31.
24 ibid.
25 Hindmarsh, 'Introducing the Terrain'.
26 Frackler, Report Finds Japan Underestimated Tsunami Danger'.
27 UBS, 'Can Nuclear Power Survive Fukushima'.
28 Suzuki, 'Managing the Fukushima Challenge'.
29 Globescan, Opposition to Nuclear Power Grows.
30 UBS, op. cit.
31 ibid.

Chapter 8: Nuclear Politics in Twenty-first Century Australia

1 Ecologically Sustainable Development Steering Committee, Summary Report on the Implementation of the National Strategy for Ecologically Sustainable Development.
2 Miller, 'The claims are exaggerated'.
3 Gratton, 'Turnbull and PM at loggerheads on Kyoto'.
4 Howard, Press Conference Parliament House.
5 ibid.
6 *Sydney Morning Herald*, 'Nuclear Power in Australia within 10 years'.
7 Department of Prime Minister and Cabinet, *Uranium Mining, Processing and Nuclear* Energy, quoted in Lowe, *Reaction Time*, p. 34.
8 Lowe, *Reaction Time*, p. 38.
9 ibid., p. 39.
10 House of Representatives Standing Committee on Industry and Resources, *Australia's uranium*.
11 Australian Uranium Association, Australian uranium and the nuclear fuel cycle; Austrade, Dynamic Industries.
12 Australian Nuclear Science and Technology Organisation (ANSTO), The Case for a New Reactor.
13 ANSTO, *Overview of Proposed Replacement Research Reactor*.
14 Broinowski, *Fact of Fission?*, pp. 243–47.
15 ibid., p. 245.
16 Green, Future Supply of Medical Radioisotopes in Australia.
17 Broinowski, op, cit., p. 243.
18 ANSTO, *Overview of Proposed Replacement Research Reactor*, p. 7.
19 International Atomic Energy Agency, New Members Elected to IAEA Board of Governors.
20 Broinowski, op. cit., pp. 244–45.

Chapter 9: Radioactive Waste – A Continuing Problem

1 Senate Select Committee on Uranium Mining and Milling, Uranium Mining and Milling in Australia.
2 Fox et al., *Ranger Uranium Environmental Inquiry First Report*, p. 90.

3   Nuclear Fuel Cycle Royal Commission, Nuclear Fuel Cycle Royal Commission Report, p. 73.
4   ibid., p. 73.
5   ibid., p. 74.
6   ibid., p. 75.
7   ibid., p. 76.
8   ibid., p. 79.
9   ibid., p. 80.
10   Bureau of Resource Sciences, *A Radioactive Waste Repository for Australia Site Selection Study Phase 3*.
11   Australianmap.net, Woomera – proposed nuclear dump site 1998–2004.
12   Donnelly, 'Victory for traditional owners over Muckaty Station nuclear waste dump'.
13   Pitt, $4 million in new community grant funding for regional South Australian communities.
14   South Australia Native Title Service, Barngala Speak Out.
15   ABC News, 'Senate inquiry recommends passing nuclear waste site at Napandee, South Australia'.
16   Pitt, op. cit.
17   Broinowski, *Fact or Fission?*, p. 247.
18   ibid., pp. 247–48.
19   ibid., p. 249.

Chapter 10: Waste and the SA Royal Commission

1   Broinowski, *Fact or Fission?*, p. 252.
2   ibid., p. 252.
3   Macken, 'Out then back'.
4   Nuclear Fuel Cycle Royal Commission, Report.
5   No Dump Alliance, Standing Strong 2015–2017.
6   Nuclear Fuel Cycle Royal Commission, Nuclear Fuel Cycle Royal Commission Report, p. xiii.
7   ibid., pp. xiv–xvi.
8   ibid., p. 1.
9   International Energy Agency, *2020 Global overview*.
10   Nuclear Fuel Cycle Royal Commission, op. cit. p. 4.
11   ibid., p. xv.
12   ibid., p. 111.
13   ibid., p. xv.
14   Australian Academy of Technology and Engineering (ATSE), Submission to the House of Representatives Committee on Environment and Energy Inquiry into the Pre-requisites for Nuclear Energy.
15   Nuclear Fuel Cycle Royal Commission, op. cit., pp. 29–42.
16   Colless, 'Tonkin bids for first N-plant'.
17   Nuclear Fuel Cycle Royal Commission, op. cit., p. 31.
18   ibid., p. 31.
19   ibid., p. 31.
20   ibid., p. 31.
21   ibid., p. 81.

22   ibid., pp. 83–85.
23   ibid., pp. 91–93.
24   ibid., p. 93.
25   ibid., pp. 99–106.
26   ibid., pp. 290–304.
27   ibid., pp. 172–73.
28   Green, 'No Way!'.

Chapter 11: Nuclear Politics in 2021

1   Martin and Cooper, 'How an American Tech Icon Bet on Nuclear – and Lost'.
2   Ambrose, 'Hinkley Point nuclear plant building costs rise by up to £2.9bn'.
3   Graham et al., *GenCost 2018.*
4   Graham et al., *GenCost 2020–21.*
5   Schneider and Froggatt, The World Nuclear Status Report 2020.
6   Parkinson, 'New CSIRO, AEMO study confirms wind, solar and storage beat coal, gas and nuclear'.
7   Blakers et al., Pumped hydro storage and 100% renewable electricity; Baldwin et al., 'At its current rate, Australia is on target for 50% renewable electricity by 2025'.
8   Diesendorf and Elliston, 'The feasibility of 100% renewable electricity systems'.
9   Deane, 'The EU is considering carbon tariffs on Australian exports'.
10   Deutsche Welle, 'Every euro invested in nuclear power makes the climate crisis worse'.
11   Energy Efficiency and Greenhouse Working Group, *Towards a National Framework for Energy Efficiency.*
12   Climate Council, Summer Stats.
13   Finkel, *Getting to Zero.*
14   Hughes, 'Does Australia need Nuclear-Powered Submarines and a Nuclear-Power Sector?'
15   Independent and Peaceful Australia Network, Background to the Inquiry.

Chapter 12: Australia and Nuclear Weapons

1   Bulletin of the Atomic Scientists, This is your COVID wake-up call; Lowe, 'What is the Doomsday Clock and why should we keep track of the time?'
2   Broinowski, *Fact or Fission?*, pp. 206–207.
3   ibid., p. 206.
4   ibid., pp. 207–10.
5   ibid., p. 223.
6   ElBaradei, Director General's Statement, 2005 Review Conference of the Treaty on Non-Proliferation of Nuclear Weapons.
7   Broinowski, op. cit., pp. 227–28.
8   ibid., pp. 279–81.
9   United Nations, Treaty on the prohibition of nuclear weapons.
10   White, *How to Defend Australia.*
11   Falk, *Taking Australia Off the Map*, pp. 162–66.
12   ibid., p. 223.

13 Independent and Peaceful Australia Network (IPAN), *Warmongering condemned*.
14 Toohey, *Secret*, p. 294.
15 Milliken, 'To Russia With Love'.
16 World Nuclear Association, *Australia's Uranium*.
17 Department of Foreign Affairs and Trade, *Australia's network of nuclear cooperation agreements*.
18 Fox et al., *Ranger Uranium Environmental Inquiry First Report*, p. 147.
19 Broinowski, op cit., p. 262.
20 Alvarez, *Report– the Global Crisis of Nuclear Waste*.
21 Green and Noonan, 'Australian uranium fuelled Fukushima'.
22 Broinowski, op. cit., p. 262.
23 Green & Noonan, 'Australian uranium fuelled Fukushima'.
24 Broinowski, op. cit., p. 282.
25 Fox et al., op. cit., pp. 185–86.
26 ibid., p. 187.

Conclusion

1 Tombs, 'Nuclear power and the public good'.
2 *Guardian*, 'The Guardian Windscale'.
3 Wynne, *Rationality and Ritual*, p. 176.
4 Ainsworth and Carrington, 'BSE disaster'.
5 Oreskes and Conway, *Merchants of Doubt*.
6 Bilinovich, 'Averting nuclear apocalypse'.
7 Toohey, *Secret*, p. 301.
8 Fraser, 'Let's not wake up to the sight of a city in ashes'.

Appendix

1 Australian Radiation Protections and Nuclear Safety Agency (ARPANSA), *CT scans for children*.
2 ARPANSA, *Patients receiving less radiation from CT scans*.
3 Pope, *Living with Radiation*, p. 144.
4 ARPANSA, *Radiation Protection and Radioactive Waste Management in Mining and Mineral Processing*.
5 Energy Resources Australia, *ERA announces Closure Plan for Ranger Mine*.
6 Broinowski, *Fact or Fission?*, p. 253.

BIBLIOGRAPHY

ABC News, 'Senate inquiry recommends passing nuclear waste site at Napandee, South Australia', 15 September 2020, accessed at https://www.abc.net.au/news/2020-09-15/senate-committee-recommends-nuclear-waste-facility-at-napandee/12658266.

Abdulkarim, A.J, and N.J.D. Lucas, 'Economies of scale in electricity generation in the United Kingdom', *International Journal of Energy Research*, vol. 1, pp. 223–32, 1977, accessed at https://onlinelibrary.wiley.com/doi/abs/10.1002/er.4440010303.

Ainsworth, C., and D. Carrington, 'BSE disaster: the history', *New Scientist*, 2000, accessed at https://www.newscientist.com/article/dn91-bse-disaster-the-history/.

Alvarez, R., H. Ban, M. Goldstick, B. Laponche, P. Roche and B. Thuillier, *Report – The Global Crisis of Nuclear Waste*, Greenpeace France, 2019, accessed at https://www.greenpeace.fr/report-the-global-crisis-of-nuclear-waste/.

Amalgamated Metal Workers & Shipwrights Union, Australian Railways Union, Building Workers' Industrial Union and Federated Miscellaneous Workers Union, *The Uranium Debate: A Trade Union Contribution*, Globe Press, Fitzroy, 1977.

Ambrose J., 'Hinkley Point nuclear plant building costs rise by up to £2.9bn', *The Guardian*, 26 September 2021, accessed at https://www.theguardian.com/uk-news/2019/sep/25/hinkley-point-nuclear-plant-to-run-29m-over-budget.

AP News, 'Hot weather forces 4 French nuclear reactors to shut down', 5 August 2018, accessed at https://apnews.com/article/2fef5c6a978b4fd5a79596020f0bac8a.

Austrade, Dynamic Industries, Why Australia Benchmark Report, 2021, accessed at https://www.austrade.gov.au/benchmark-report/dynamic-industries.

Australian Academy of Technology and Engineering (ATSE), Submission to the House of Representatives Committee on Environment and Energy Inquiry into the Pre-requisites for Nuclear Energy, September 2019, accessed at https://www.atse.org.au/wp-content/uploads/2019/09/Inquiry-into-the-prerequisites-for-nuclear-energy.pdf.

Australianmap.net, Woomera – proposed nuclear dump site 1998–2004, Australian Nuclear and Uranium Sites, 2012, accessed at https://nuclear.australianmap.net/woomera/.

Australian Nuclear Science and Technology Organisation (ANSTO), The Case for a New Reactor, unpublished manuscript, 1991.

—— *Overview of Proposed Replacement Research Reactor*, PPK Environment & Infrastructure, Sydney, 1997.

Australian Radiation Protections and Nuclear Safety Agency (ARPANSA), CT scans for children: information for referrers, 2016, accessed at https://www.arpansa.gov.au/understanding-radiation/radiation-sources/more-radiation-sources/ct-imaging-and-children-referrers.

—— *Exposure of Humans to Ionizing Radiation for Research Purposes*, Radiation Protection Series No. 8 ARPANSA, Yallambie, 2005.

—— *National Directory of Radiation Protection – Edition 1.0*, Radiation Protection Series No. 6, ARPANSA, Yallambie, 2014.

—— Patients receiving less radiation from CT scans, media release, 2020, formerly available from https://www.arpansa.gov.au/news/patients-receiving-less-radiation-ct-scan.

—— *Radiation Protection and Radioactive Waste Management in Mining and Mineral Processing*, Radiation Protection Series No. 9, ARPANSA, Yallambie, 2005.

—— *Recommendations for Limiting Exposure to Ionizing Radiation … and National Standard for Limiting Occupational Exposure to Ionizing Radiation*, Radiation Protection Series No. 1, ARPANSA, Yallambie, 2002.

Australian Science, Technology and Engineering Council (ASTEC), *Australia's Role in the Nuclear Fuel Cycle*, Australian Government Publishing Service, Canberra, 1984.

—— *Nuclear Science and Technology in Australia: A report to the Prime Minister*, Australian Government Publishing Service, Canberra, 1985.

Australian, The, 'Editorial', p. 6, 8 November 1978.

Australian Uranium Association, Australia's uranium and the nuclear fuel cycle, Information Pack, May 2011, accessed at https://www.parliament.vic.gov.au/images/stories/committees/edic/greenfields_mineral_exploration/subs/29_-_Australian_Uranium_Association_-_Attachment_2.pdf.

Australian Uranium Producers' Forum, *Uranium Mining and Nuclear Power*, no publisher stated, undated.

Baldwin, K., A. Blakers and M. Stock, 'At its current rate, Australia is on track for 50% renewable electricity in 2025', *Social Policy Connections*, 10 September 2018, accessed at https://www.socialpolicyconnections.com.au/?p=13910.

Bertell, R., *No Immediate Danger: Prognosis for a Radioactive Earth*, The Women's Press, London, 1985.

Bilinovich, B., 'Averting nuclear apocalypse', *The Observer*, 15 March 2021, accessed at https://observer.case.edu/bilinovich-averting-nuclear-apocalypse/.

Blakers, A., B. Lu & M. Stocks, Pumped hydro storage and 100% renewable electricity, presentation to the Hawaii World Conference on Photovoltaic Energy Conversion, Australian National University, 2018, accessed at https://www.dropbox.com/s/tkratcm2j8rexp5/Blakers%20100%25.pdf?dl=0.

Broinowski, R., *Fact or Fission?: The truth about Australia's nuclear ambitions*, Scribe Publications, Carlton North, 2003.

Brook, B., & I. Lowe, *Why vs Why: Nuclear Power*, Pantera Press, Sydney, 2010.

Bulletin of the Atomic Scientists, This is your COVID wake-up call: It is 100 seconds to midnight, 2021, accessed at https://thebulletin.org/doomsday-clock/current-time/.

Bureau of Resource Sciences, *A Radioactive Waste Repository for Australia Site Selection Study Phase 3, Regional Assessment*, Commonwealth of Australia, Canberra, 1997.

Byrne, L., 'Wilson's White Heat', Fabian Society, UK, 11 March 2016, accessed at https://fabians.org.uk/wilsons-white-heat/.

Caldicott, H., *Nuclear Power is Not the Answer to Global Warming or Anything Else*, Melbourne University Press, Carlton, 2006.

Cawte, A., *Atomic Australia: 1944–1990*, New South Wales University Press, Kensington, 1992.

Chapman, P., *Fuel's Paradise: Energy options for Britain*, Penguin Books, Harmondsworth, 1975.

Christoff, P., 'The Nuclear Disarmament Party', *Arena*, vol. 70, pp. 9–29, 1985.

Climate Council, Summer Stats: Solar outshone gas, 16 March 2021, accessed at
https://www.climatecouncil.org.au/resources/summer-gas-stats-2020-21/.

Cockburn, S., and D. Ellyard, *Oliphant: The life and times of Sir Mark Oliphant*, Axiom
Books, Kent Town, 1981.

Colless, M., 'Tonkin bids for first N-plant', *The Financial Australian*, 23 July 1980.

Collingridge, D., *Technology in the Policy Process: The control of nuclear power*, St Martin's,
New York, 1983.

—— *The Social Control of Technology*, Frances Pinter, London, 1980.

Coopers & Lybrand Consultants, Financial and Economic Evaluation of Proposed
New Reactor, unpublished manuscript, 1991.

Deane, F., 'The EU is considering carbon tariffs on Australian exports: Is that legal?',
The Conversation, 26 March 2021, accessed at https://theconversation.com/the-
european-union-wants-to-impose-carbon-tariffs-on-australian-exports-is-that-
legal-156946.

Department of Education, Science and Training, *Safe Storage of Radioactive Waste*,
Commonwealth of Australia, Canberra, 2002.

Department of Foreign Affairs and Trade, Australia's network of nuclear
cooperation agreements, 2020, accessed at https://www.dfat.gov.au/
international-relations/security/non-proliferation-disarmament-arms-
control/policies-agreements-treaties/nuclear-cooperation-agreements/Pages/
australias-network-of-nuclear-cooperation-agreements.

Department of Primary Industries and Energy, *National Radioactive Waste Repository
Site Selection Study Phase 2*, Australian Government Publishing Service, Canberra,
1995.

Department of Prime Minister and Cabinet, *Uranium Mining, Processing and Nuclear
Energy: Opportunities for Australia?* Report of the Uranium Mining, Processing
and Nuclear Energy Review, 2006, formerly available at
https://www.pmc.gov.au/umpner/docs/nuclear_report.pdf.

Deutsche Welle, 'Every euro invested in nuclear power makes the climate crisis worse',
16 March 2021, accessed at https://www.dw.com/en/nuclear-climate-mycle
-schneider-renewables-fukushima/a-56712368.

Diesendorf, M., and B. Elliston, 'The feasibility of 100% renewable electricity systems:
A response to critics', *Renewable and Sustainable Energy Reviews*, vol. 93, pp. 318–
30, October 2018, accessed at https://www.sciencedirect.com/science/article/abs/
pii/S1364032118303897.

Donelly, B., 'Victory for traditional owners over Muckaty Station nuclear waste dump',
The Sydney Morning Herald, 19 June 2014, accessed at https://www.smh.com.au/
politics/federal/victory-for-traditional-owners-over-muckaty-station-nuclear-
waste-dump-20140619-zsedf.html.

Douglas, R., et al., *W(h)ither Australia? Surviving and Thriving in a Mega-Threatened
World*, Commission for the Human Future, Canberra, 20 January 2021, accessed
at https://www.dropbox.com/s/teso10qb4nvym6o/Final%20%20Roundtable%20
Report.pdf?dl=0.

Ecologically Sustainable Development Steering Committee, Summary Report on
the Implementation of the National Strategy for Ecologically Sustainable
Development, Commonwealth of Australia, 1994.

ElBaradei, M., Director General's Statement, 2005 Review Conference of the Treaty
on Non-Proliferation of Nuclear Weapons, International Atomic Energy Agency,

Vienna, 2005, accessed at https://www.iaea.org/newscenter/statements/treaty-non-proliferation-nuclear-weapons.

Elliott, M. (ed.), *Ground for Concern: Australia's Uranium and Human Survival*, Penguin books, Harmondsworth, 1977.

Energy Efficiency and Greenhouse Working Group, *Towards a National Framework for Energy Efficiency: Issues and Challenges*, Discussion Paper, 2003, accessed at https://www.airah.org.au/Content_Files/Advocacy/Sustainability/NFEE_Stage1_DiscussionPaper-2004.pdf.

Energy Resources Australia, ERA releases Closure Plan for Ranger Mine, media release 2021, accessed at https://www.energyres.com.au/media/era-releases-closure-plan-for-ranger-mine/.

Environment Australia, *Environment Assessment Report: Proposed Replacement Nuclear Research Reactor at Lucas Heights*, Department of the Environment and Heritage, Canberra, 1999.

Falk, J., *Taking Australia Off the Map: Facing the Threat of Nuclear War*, Penguin Books, Ringwood, 1983.

Findlay, T., *The Future of Nuclear Energy to 2030 and Its Implications for Safety, Security and Non-proliferation*, Centre for International Governance Innovation, Waterloo, Ontario, 2010, accessed at https://www.cigionline.org/sites/default/files/part_1.pdf.

Finkel, A., *Getting to Zero: Australia's Energy Transition*, Quarterly Essay No. 81, Black Inc., 2021.

Fox, R.W., G.G. Kelleher and C.B. Kerr, *Ranger Uranium Environmental Inquiry First Report*, Australian Government Printing Service, Canberra, 1976.

Frackler, M., Report Finds Japan Underestimated Tsunami Danger, *The New York Times*, 1 June 2011, accessed at http://www.nytimes.com/2011/06/02/world/asia/02japan.html?_r=2&ref=world&.

Fraser, J.M., Answers to Questions: Nuclear waste disposal (Question No. 5418), *Hansard*, 19 August 1980 (question asked 20 February 1980), accessed at https://parlinfo.aph.gov.au/parlInfo/search/display/display.w3p;db=HANSARD80;id=hansard80%2Fhansardr80%2F1980-08-19%2F0211;query=Id%3A%22hansard80%2Fhansardr80%2F1980-08-19%2F0325%22.

—— 'Let's not wake up to the sight of a city in ashes', *The Australian* 26 January 2014.

Globescan, Opposition to Nuclear Power Grows: Global Poll, 15 November 2011, accessed at https://globescan.com/opposition-to-nuclear-energy-grows-global-poll/.

Graham, P.W., J. Hayward, J. Foster and L. Havas (2021) *GenCost 2020–21: Consultation draft*, CSIRO, 2021, accessed at https://publications.csiro.au/publications/publication/PIcsiro:EP208181/SQGenCost2020-21/RP1/RS25/RORECENT/STsearch-by-keyword/LISEA/RI1/RT1.

Graham, P.W., J. Hayward, J. Foster, O. Story and L. Havas, *GenCost 2018: Updated projections of electricity generation technology costs*, CSIRO, December 2018, accessed at https://publications.csiro.au/rpr/download?pid=csiro:EP189502&dsid=DS1.

Gratton, M., 'Turnbull and PM at loggerheads on Kyoto', *The Sydney Morning Herald*, 28 October 2007, accessed at https://www.smh.com.au/politics/federal/turnbull-and-pm-at-loggerheads-on-kyoto-20071028-gdrg9v.html.

Green, J., Future Supply of Medical Radioisotopes in Australia: Do We Need A New Reactor?, extracts from PhD thesis, University of Wollongong, 1999.

—— 'No way! South Australians reject international nuclear waste dump', *The Ecologist* 9 November 2016, accessed at https://theecologist.org/2016/nov/09/no-way-south-australians-reject-international-nuclear-waste-dump.

—— The push for nuclear weapons in Australia 1950s–1970s, Friends of the Earth Australia, 1999, accessed at https://www.foe.org.au/anti-nuclear/issues/oz/ozbombs.

Green, J., and D. Noonan, 'Australian uranium fuelled Fukushima', *The Ecologist*, 9 March 2021, accessed at https://theecologist.org/2021/mar/09/australian-uranium-fuelled-fukushima.

Guardian, The, 'The Guardian Windscale: A summary of the evidence and the Argument', Guardian Newspapers, London, 1977.

Hall, T., *Nuclear Politics: The history of nuclear power in Britain*, Penguin Books, Harmondsworth, 1986.

Hara, T., 'Social Structure and Nuclear Power Siting Problems Revealed' in R. Hindmarsh (ed.), *Nuclear Disaster at Fukushima Daiichi*, op. cit., pp. 41–56, 2013.

Haynes, P., and M. Bojcun, *The Chernobyl Disaster*, Hogarth Press, London, 1988.

Hindmarsh, R., 'Introducing the Terrain' in R. Hindmarsh (ed.), *Nuclear Disaster at Fukushima Daiichi*, op. cit., pp. 1–21, 2013.

—— (ed.), *Nuclear Disaster at Fukushima Daiichi: Social, Political and Environment Issues*, Routledge, New York & Abingdon, 2013.

Hore-Lacy, I., and R. Hubery, *Nuclear Electricity: An Australian perspective*, Australian Mining Industry Council, Dickson, 1989.

House of Representatives Standing Committee on Industry and Resources, *Australia's uranium: Greenhouse friendly fuel for an energy-hungry world*, November 2006, accessed at https://www.aph.gov.au/binaries/house/committee/isr/uranium/report/fullreport.pdf.

Howard, J., Press Conference Parliament House, Canberra, 6 June 2006, accessed at https://pmtranscripts.pmc.gov.au/release/transcript-22314.

Hoyle, F., *Energy or Extinction? The case for nuclear energy*, Heinemann, London, 1977.

Hughes, L., 'Does Australia need Nuclear-Powered Submarines and a Nuclear-Power Sector?', Future Directions International, Perth, 2021.

Independent and Peaceful Australia Network (IPAN), Background to the Inquiry, 2021, accessed at https://independentpeacefulaustralia.com.au/background/.

—— Warmongering condemned, media release 15 March 2021, accessed at https://ipan.org.au/warmongering-condemned-15th-march-2021/.

International Atomic Energy Agency, New Members Elected to IAEA Board of Governors, media release, 24 September 2020, accessed at https://www.iaea.org/newscenter/news/new-members-elected-to-iaea-board-of-governors-24-september-2020.

International Commission on Radiological Protection (ICRP), 2007 Recommendations of the International Commission on Radiological Protection, *Annals of the ICRP*, Publication 103, Elsevier, 2007.

International Energy Agency, *2020 Global overview: Capacity, supply and emissions*, 2020, accessed at https://www.iea.org/reports/electricity-market-report-december-2020/2020-global-overview-capacity-supply-and-emissions.

Leatherdale, D., 'Windscale Piles: Cockcroft's Follies avoided nuclear disaster', BBC News, 4 November 2014, accessed at https://www.bbc.com/news/uk-england-cumbria-29803990.

Lowe, I., 'Blinded by Science: Labor and Slatyer", *Arena*, no. 69, pp. 133–150, 1984.

—— *Living in the Greenhouse*, Scribe, Newham, 1969.

—— *Reaction Time: Climate change and the nuclear option*, Quarterly Essay No. 27, Black Inc, 2007.

—— 'What is the Doomsday Clock and why should we keep track of the time?' *The Conversation*, 27 January 2017, accessed at https://theconversation.com/what-is-the-doomsday-clock-and-why-should-we-keep-track-of-the-time-71990.

Macintyre, D., 'How the miners' strike of 1984–85 changed Britain for ever', *New Statesman*, 16 June 2014, accessed at https://www.newstatesman.com/politics/2014/06/how-miners-strike-1984-85-changed-britain-ever.

Macken, J., 'Out, then back: The big N-plan', *Financial Review*, 6 June 2006, accessed at https://www.afr.com/markets/commodities/out-then-back-the-big-n-plan-2006 0607-j7aip.

Martin, B., *Nuclear Knights*, Rupert Public Interest Movement, Canberra, 1980.

Martin, C., and C. Cooper, 'How an American Tech Icon Bet on Nuclear – and Lost', *Bloomberg*, 31 March 2017, accessed at https://www.bloomberg.com/news/articles/2017-03-29/how-an-american-tech-icon-bet-on-nuclear-and-lost-its-way.

Melbourne, A.J., Australia's Radium Legacy Waste, presentation to Radiation Health and Safety Advisory Council, ARPANSA, Yallambie, 2009.

Middleton, S., 'No Uranium Mining in SA', *The News*, 5 February 1979, p. 1, cited in Woollacott, A., *Don Dunstan: The visionary politician who changed Australia*, Allen & Unwin Sydney, 2019.

Miller, N., '"The claims are exaggerated": John Howard rejects predictions of global warming catastrophe', *Sydney Morning Herald*, 6 November 2013, accessed at https://www.smh.com.au/politics/federal/the-claims-are-exaggerated-john-howard-rejects-predictions-of-global-warming-catastrophe-20131106-2wzza.html.

Milliken, R., *No Conceivable Injury: The story of Britain and Australia's atomic cover-up*, Penguin Books, Ringwood, 1986.

—— 'To Russia With Love', *National Times*, 7–13 January 1982.

Morone, J., and E.J. Woodhouse, *The Demise of Nuclear Energy?: Lessons for democratic control of technology*, Yale University Press, New Haven & London, 1987.

Moss, N., *Men Who Play God: The story of the hydrogen bomb*, Harper & Row, New York, 1968.

Moyal, A.M., 'The Australian Atomic Energy Commission: A case study in Australian science and government', *Search*, vol. 6, p. 9, 1975.

Murakami, S., and A. Sheldrick, 'Climbing without a map: Japan's nuclear clean-up has no end in sight', Swissinfo, 12 March 2021, accessed at https://www.swissinfo.ch/eng/climbing-without-a-map--japan-s-nuclear-clean-up-has-no-end-in-sight/46442618.

Myhra, S., 'Three Mile Island and Implications for Australian Uranium Mining', *Search*, vol. 10, pp. 350–55, 1979.

Nathan, S., 'Energy and Thatcher: A tangled legacy', *The Engineer*, 10 April 2013, accessed at https://www.theengineer.co.uk/energy-and-thatcher-a-tangled-legacy/.

No Dump Alliance, *Standing Strong 2015–2017*, Conservation Council of SA, Adelaide, 2018.

Nuclear Fuel Cycle Royal Commission, *Nuclear Fuel Cycle Royal Commission Report*, Government of South Australia, Adelaide, 2016.

Oreskes, N., and E.M. Conway, *Merchants of Doubt: How a handful of scientists obscured the truth on issues from tobacco smoke to global warning*, Bloomsbury, New York, 2010.

Parkinson, G., 'New CSIRO, AEMO study confirms wind, solar and storage beat coal, gas and nuclear', *RenewEconomy*, 6 February 2020, accessed at https://reneweconomy.com.au/new-csiro-aemo-study-confirms-wind-solar-and-storage-beat-coal-gas-and-nuclear-57530/.

Patterson, W., *Nuclear Power*, Penguin Books, Harmondsworth, 1976.

Penney, W., B.F. Schonland, J. M. Kay, J. Diamond and D.E.H. Peirson, Report on the accident at Windscale No. 1 Pile on 10 October 1957, reprinted in September 2017, *Journal of Radiological Protection*, vol. 37, no. 3, pp. 780–96, accessed at https://www.researchgate.net/publication/319389458_Report_on_the_accident_at_Windscale_No_1_Pile_on_10_October_1957.

Pitt, K., $4 million in new community grant funding for regional South Australian communities, media release, 2021, accessed at https://www.minister.industry.gov.au/ministers/pitt/media-releases/4-million-new-community-grant-funding-regional-south-australian.

Pollard, R.D. (ed.), *The Nugget File: excerpts from the government's special file on nuclear power plant accidents and safety defects, obtained by the Union of Concerned Scientists under the Freedom of information act*, Union of Concerned Scientists, John Wiley & Sons, New York, 1979.

Pollitt, M.G., 'UK final electricity demand by sector 1960–2009' in *The Economics of Energy (and Electricity) Demand*, Researchgate file, April 2011, accessed at https://www.researchgate.net/publication/241753448_The_Economics_of_Energy_and_Electricity_Demand/figures?lo=1.

Pope, C., *Living with Radiation*, Inspiring Publishers, Calwell, ACT, 2018.

PPK Environment & Infrastructure, *Overview of Proposed Replacement Nuclear Research Reactor*, ANSTO, Sydney, 1997.

Pringle, P., and J. Spigelman, *The Nuclear Barons*, Sphere Books, London, 1982.

Ringwood, A.E., S.E. Kesson, N.G. Ware, W. Hibberson and A. Major, 'Immobilisation of high level nuclear reactor wastes in SYNROC', *Nature*, vol. 278, pp. 219–23, 1979.

Sabbagh, D., 'Cap on Trident nuclear warhead stockpile to rise by more than 40%', *The Guardian*, 16 March 2021, accessed at https://www.theguardian.com/uk-news/2021/mar/15/cap-on-trident-nuclear-warhead-stockpile-to-rise-by-more-than-40.

Scarce, K., *Nuclear Fuel Cycle Royal Commission Report*, Government of SA, Adelaide, 2016.

Schneider, M., and A. Froggatt, et al., The World Nuclear Industry Status Report 2020, accessed at https://www.worldnuclearreport.org/The-World-Nuclear-Industry-Status-Report-2020-HTML.html#_idTextAnchor421.

Schroeer, D., *Science, Technology and the Nuclear Arms Race*, John Wiley & Sons, New York, 1984.

Select Panel of the Public Inquiry into Uranium, *The Report of the Public Enquiry into Uranium*, Conservation Centre, Adelaide. 1997.

Senate Select Committee on Uranium Mining and Milling, Uranium Mining and Milling in Australia: Findings, conclusions and recommendations, 1997, accessed at https://www.aph.gov.au/Parliamentary_Business/Committees/Senate/Former_Committees/uranium/report/b02.

South Australia Native Title Service, Barngala Speak Out, 2020, accessed at https://www.nativetitlesa.org/barngarla-speak-out/.

Suter, K., *Australia and the Nuclear Choice: The report of the Independent Commission of Inquiry into nuclear weapons and other consequences of Australian uranium mining*, Total Environment Centre, Sydney, 1984, accessed at https://trove.nla.gov.au/work/13371561.

Suzuki, A., 'Managing the Fukushima Challenge', *Risk Analysis*, vol. 34, Issue 7, 1240–256, 20 June 2014, accessed at https://onlinelibrary.wiley.com/doi/full/10.1111/risa.12240.

Switkowski, Z., Uranium Mining, Processing and Nuclear Energy: Opportunities for Australia?, report, also known as UMPNER or the Switkowski review, Australian Government, Canberra, 2006.

Sydney Morning Herald, The, 'Nuclear power in Australia within 10 years: Switkoski', 27 November 2006, accessed at https://www.smh.com.au/national/nuclear-power-in-australia-within-10-years-switkowski-20061127-gdowxg.html.

Titterton, E., 'World Energy Requirements and their Supply', *Search*, vol. 13, pp. 253–56, 1982.

Tombs, F., 'Nuclear power and the public good', *Atom*, vol. 255, pp. 2–8, 1978.

Toohey, B., *Secret: The making of Australia's security state*, Melbourne University Press, Carlton, 2019.

UBS, 'Can Nuclear Power Survive Fukushima', UBS Investment Research, Q-Series: Global Nuclear Power, 4 April 2011.

United Nations, Treaty on the prohibition of nuclear weapons, Office for Disarmament Affairs, accessed at https://www.un.org/disarmament/wmd/nuclear/tpnw/.

U.S. Nuclear Regulatory Commission, Reactor Safety Study WASH-1400, U.S. Government Printing Service, Washington, 1975.

Watters, R.A., and S. Chandra, *The Nuclear Power Industry: A responsible approach*, Government Printer of the Northern Territory, Darwin, 1985.

Weinberg, A.M., 'Science and Trans-Science', *Minerva*, vol. 9, pp. 220–32, 1968.

White, H., *How to Defend Australia*, Black Inc., Melbourne, 2019.

White, T.J., R.L. Segall and P.S. Turner, 'Radwaste Immobilization by Structural Modification: The crystallochemical properties of Synroc, a titanate ceramic', *Angewandte Chemie*, International Edition in English, vol. 24, no. 5, p. 357–438, May 1985, accessed at https://www.docme.su/doc/1951281/radwaste-immobilization-by-structural-modificationchthe-cr.

World Nuclear Association, *Australia's Uranium*, 2020, accessed at https://www.world-nuclear.org/information-library/country-profiles/countries-a-f/australia.aspx.

—— *Chernobyl Accident 1986*, 2020, accessed at https://www.world-nuclear.org/information-library/safety-and-security/safety-of-plants/chernobyl-accident.aspx.

—— *Fukushima Daiichi Accident*, 2021, accessed at https://www.world-nuclear.org/information-library/safety-and-security/safety-of-plants/fukushima-daiichi-accident.aspx.

Wynne, B., *Rationality and Ritual: The Windscale Inquiry and nuclear decisions in Britain*, BSHS monographs 3, British Society for the History of Science, Chalfont, St Giles, 1982.

Yamamitsu, E., 'Ten years on, Japan mourns victims of earthquake and Fukushima disaster', Reuters Wire Service, 11 March 2021, accessed at https://www.reuters.com/article/us-japan-fukushima-anniversary-idUSKBN2B301A.

INDEX

ACKNOWLEDGEMENTS

Many scientists contributed to my intellectual development and equipped me to write this book. The late professors C.J. Milner and J.C. Kelly at University of New South Wales were my early mentors, Dr David Blackburn and Professor Michael Wolfson guided me at University of York, then Professor Charles Newey gave me my first academic post at the UK Open University and encouraged me to set high standards.

My colleagues in the School of Science at Griffith University, especially those in the Science, Technology and Society group, constantly supported me and challenged my thinking. I am particularly grateful to the late Ian Henderson, a Griffith University colleague who chaired the Campaign Against Nuclear Power, and Dr John Forge, whose introduction of a course on nuclear energy, weapons and warfare forced me to get more deeply involved in these troubling issues.

My period advising Western Mining when they operated the Roxby Downs mine and my years on the Radiation Health and Safety Advisory Council gave me a much better understanding of the complex issues involved in regulating ionising radiation.

This book originated from the award of a Griffith Review Writing Fellowship, funded by the Queensland Government through Arts Queensland. That support, for which I am deeply appreciative, enabled me to write an essay entitled 'A long half-life' for Griffith Review 71, *Remaking the Balance*, before developing the argument into this book. I gratefully acknowledge both the support and the wise advice from the editor of Griffith Review, Ashley Hay.

Greg Bain, as manager of Monash University Publishing, enthusiastically encouraged me to submit the proposal for this book and then turn it into reality. Joanne Holliman did a wonderful job editing the manuscript and has improved it out of sight. Sherrey Quinn has created a very comprehensive index.

My late partner, Dr Patricia Kelly, was always metaphorically by my side as I wrote this book. It is dedicated to her memory.

None of those wise people are to blame for any errors or infelicities that have escaped their scrutiny. I take full responsibility for the end product.

ABOUT THE AUTHOR

Professor Ian Lowe AO is uniquely qualified to tell this story, following a long career in universities, research councils and advisory groups. Lowe is the author of several books, including *Living in the Hothouse* (Scribe, 2005), *A Big Fix* (Black Inc., 2005), *A Voice of Reason* (UQP, 2010), *Bigger or Better?* (UQP, 2012) and *The Lucky Country?* (UQP, 2016). He is also the author of a 2006 *Quarterly Essay* on the prospects for nuclear power in Australia, and a 'flip book' with Professor Barry Brook, giving the two sides of the argument.

CPSIA information can be obtained
at www.ICGtesting.com
Printed in the USA
LVHW110900250821
696024LV00003B/96